Behind the structure of our buildings,
Behind the functions of our bodies,
Behind the motions of nature,
Behind art and music,
Behind everyday phenomena…
  mathematics is there —
  explaining, describing & influencing.

—Theoni Pappas

Math

Stuff

by theoni pappas

— Wide World Publishing/Tetra —

Wide World Publishing/Tetra
P.O. Box 476
San Carlos, CA 94070

Printed in the United States of America.

1st Printing August 2002

ISBN: 1-884550-26-6

TABLE OF CONTENTS

TABLE OF CONTENTS

introduction

stuff *(stüf) n. 1. The material out of which something is made or formed; substance.*
v. 1. a. To pack tightly.

MATH STUFF is not a book of numbers, formulas, or computations. It's a book of ideas. The mathematics behind these ideas is discussed in general terms, and each chapter is designed to be self-contained allowing the reader to open to a topic at random. Many of the ideas presented are on the cutting edge. Their evolution and impact on us and the universe will be phenomenal.

Mathematical stuff can deal with anything from abstract concepts to fiddlehead ferns, from a number to a numeral, from software to the nuts and bolts of a computer. Stuff is the layperson's unpretentious word which has no limits or boundaries. Math stuff fills both conscious and unconscious space. Often we're oblivious to math stuff, just as we're oblivious to atoms and molecules. When crossing a bridge, most of us never think about the multiple equations and number crunching that went into the bridge's design and construction. When hiking or gardening, we don't think about the mathematics behind rock formations or plant forms. Or when flipping on the TV or using a cell phone, the mathematics of wave theory is farthest from our minds. Yet, math stuff is there describing, explaining, and influencing our lives.

Most think mathematics involves only complicated equations, graphs, diagrams, complex formulas and theorems. Many are unaware that deciding what to cook for dinner, which road to take at a fork, or prioritizing tasks are indeed forms of mathematical problem solving. Mathematics entails the discovery of patterns, be they patterns in the shape of leaves or patterns of behavior. As Ms. Fibonacci said in Jon Scieszka book MATH CURSE *"You know, you can think of almost everything as a math problem."*

For many mathematics conjures an image of a subject that is cold, sterile and elitist. Nothing could be further from the truth. Whether you are adjusting the settings of a camera, discussing politics, doing financial planning, or just talking about the weather, mathematics is there behind the scenes —organizing — describing —directing —predicting —problem solving.

Hopefully MATH STUFF will help you realize that mathematics is truly the *"stuff* that dreams are made of".

I am one and two,
1/2 and 29/30,
0.03 and 6.3333,
-8 and e,
a million and a googol,
7 and π.

I am i and 5+3i.

I am nothing and zero.
I am the set of all numbers.
I am the empty set.

I am adding and subtracting,
multiplying and dividing.

I am a quadratic equation,
a polynomial,
a coefficient,
a power and exponent.

I am squares, fractals…
I am a point,
a line, a plane.
I am space.

I am the pattern on a tortoise shell,
a spider's web,
the shape of a leaf.

I am the sound of music,
the crest of a wave.

I am the curve of a shark's fin,
the arc of a pendulum
and a unit of time.

I am unreal worlds.

I am order.

I am chaos.

I am MATHEMATICS.

—*Theoni Pappas*

the pea counts
mathematics behind your genes

The mathematical story of genetics began in the late 1850s when some of the mysteries of heredity first came to light in the garden of a monastery in Moravia. Here a lone monk named Gregor Mendel(1822-1884) avidly tended the pea plants in his garden. His interest in these plants was not aesthetic. He was conducting experiments whose results would launch the science of genetics. How could the color, the shape, and the size of a common garden pea play such a significant role in the secrets of heredity? The answer lies in how Mendel conducted his experiments. Sensing there must exist some order to how traits are passed on from one generation to the next and how traits reappear after generations have passed without their appearance, Mendel looked for and

Rendition of a portrait of Mendel

Mendel introduced the terms *dominant* and *recessive* traits in 1865. A dominant trait's characteristic, (e.g. *B*, black hair) is visible even when it is crossed with a recessive trait (e.g. *r* red hair)*Br* appears as black hair. On the other hand, a recessive trait characteristic is visible when it's the sole trait *(rr)*.

found answers in patterns revealed in the numbers and types of characteristics he tracked, counted, and tabulated over the years. He knew that the larger his sample the more accurate his observations and conclusions would be. In those days, without computers and other instruments his work was labor intensive as he manually gathered, sorted, counted, labeled and recorded. "By the time Mendel was done with his succession of crosses, recrosses, and backcrosses, he must have counted a total of more than 10,000 plants, 40,000 blossoms, and a staggering 300,000 peas."[1] Algebra, combination theory, and probability were his tools for interpretation.

What did he discover? He uncovered ratios — ratios that revealed how *dominant* and *recessive* characteristics (now called genes) are passed on from one generation to the next, and how these traits skip generations and resurface. In addition to *inventing a notation system* to keep track and interpret his results, *he interpreted his results using algebra* and adapted the expression of the binomial $(a+b)^2$.

Mendel used the ordinary garden pea such as these for his study.

He expressed dominant traits with capital letters and recessive with lower case. After analyzing his work he discovered that the algebraic binomial $(a+b)^2=a^2+2ab+b^2$ actually surfaced in his results. By labeling the *dominant trait* **A** and **a** the *recessive*, the data from his pea experiment revealed the ratio 1:2:1, and he noticed the numbers of this ratio manifested themselves as coefficients of the math-

The algebraic product of
    (A+2Aa+a)(B+2Bb+b) is
AB+2ABb+Ab+2AaB+4AaBb+2A ab+aB+2aBb+ab. Notice in nine of the terms the dominant traits **AB** appear: AB+2ABb+ 2AaB +4AaBb. In three terms the **A** is dominant & **b** recessive: Ab+2Aab. In three terms the **B** is dominant and **a** recessive: aB+2aBb. And finally, in only one are both **a** and **b** recessive: ab. The ratio 9:3:3:1 was worked out. after Mendel's work was rediscovered in the early 1900s.

ematical expression for $(A+a)(A+a) \longrightarrow 1A^2+2Aa+1a^2$. Finally, he wrote his result as $A+2Aa+a$, where a single **A** and **a** were understood to represent **AA** and **aa**. In addition, he denoted the hybrid as **Aa**, and noticed that the expression $A+2Aa+a$ actually meant the ratio 3:1 *when considering only the appearance of the pea*, since **AA** and **2Aa** *exhibit the dominant A trait in appearance* and only **aa** *exhibit the recessive trait*. His experiments were meticulous and well organized. Along with his algebraic notation and ratios he looked to combination theory to explore the possible arrangements of three traits occurring. Indeed, he went further and crossed two hybrids $(A+2Aa+a)(B+2Bb+b)$ and other crosses. Mendel's work illustrated that these units of heredity, (genes), reside in an organism's cells, and are passed on from one generation to the next according to constant ratios.

Although his phenomenal work was ground breaking, he died before he gained recognition for its impact on genetics. Fortunately his complete work, which he had preserved in lecture form, was published in 1866 in the Brünn Society's periodical Proceedings. Mendel obtained copies of the article and distributed them to prominent scientists of the time, but some recipients never even bothered to cut open its pages let alone read it. Sixteen years after Mendel's death, his work was rediscovered. Some scientists who were doing similar work came across his article, and realized it as a major scientific contribution. In fact, some had replicated similar experiments which reconfirmed and gave additional credence to Mendel's findings. Mendel's scientific work and tedious experiments were an amazing feat, especially without the use of modern technology.

The human genome project

In June 2000, 140 years after Mendel's experiments, another body of work, the Human Genome Project's (HGP) rough draft, was completed. Its result is a list of 3 billion letters composed of only four letters — A,T,G and C— appearing with no breaks in one enormously long sequence. A,T,G and C stand for four chemicals—adenine, guanine, thymine and cytosine — called *bases*. These bases occur in varying combinations, and it is these bases which form what is known as DNA. An organism's genome lists all the DNA that resides in the organism's cell on its 6. DNA can be thought of as a blueprint or a recipe for creating a particular organism — be it animals, bacteria, viruses, insects — with its unique traits. Essentially, DNA harbors the information for creating and maintaining the life of that organism. The human genome's

DNA(deoxyribonucleic acid) is made of nucleic acids with the deoxyribose sugar, while RNA(ribonucleic acid) is composed of nucleic acids with the ribose sugar. Nucleic acids are formed from nucleotides—made from a sugar, a phosphate and a base (there are five kinds of bases which are referred to by the symbols A,C,G,T,U which stand for adenine, cytosine, guanine, thymine, and uracil). In DNA, only the four bases A,T,C,G appear. Because of their molecular structure, each base can pair only with a specific base — A with T and C with G — making two strings of bases linked by the paired

Nucleus
Chromosomes
Separating strands of parent DNA
Daughter helix
Thymine
Deoxyribose
Daughter helix
Hydrogen
Carbon
Oxygen
Phosphate
A single nucleotide

Courtesy of U.S. Dept. of Energy
Human Genome Program
www.ornl.gov/hgmis

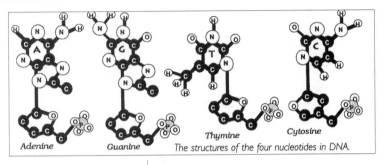

Adenine   Guanine   Thymine   Cytosine

The structures of the four nucleotides in DNA.

bases . If the bases of one side of a DNA fragment are identified, then its opposite side bases are also known because of the unique way the bases pair with one another.

For example, if a fragment's strand contains

...ACCGT... the opposite corresponding strand's fragment must

be...TGGCA...

On the other hand, RNA uses only the bases A,C,G and U, and it is composed of a single strand which is much shorter than DNA.

Chromosomes can be seen using a light microscope. Special dye techniques reveal light and dark bands indicating concentrations of a chromosome's various bases. Analyzing the sizes and patterns of these bands distin-

This histogram reflects the distribution and densities of genes along human chromosome 8. Courtesy of U.S. Dept. of Energy *Human Genome Program* www.ornl.gov/hg mis.

billion bases are arranged along its various chromosomes, and hidden among this list are some 30,000 to 40,000 genes. Imagine if the 3,000,000,000 letters were listed in book form — a human genome book. Such a book would resemble about a 200,000 page telephone book, and would take over 47 years to read if one read two letters per second.

What do scientists do with such a list? Hunting and identifying genes is one of the primary focuses of genome projects even though a little less than 2% of the DNA information of the human genome are genes. Only a very few types of DNA abnormalities, such as Down syndrome, can be detected by microscope. Most abnormalities can only be detected by focusing on changes in a gene's bases or in bases of non-gene DNA. Some genes contain information on how to make *proteins*, while other genes encode forms of RNA to operate the

6

cell by having proteins called enzymes and other molecules do the work. Locating genes along the thousands and thousands of listed bases mingled within all the non-gene information is very tricky. There are strings of DNA bases either separating genes or within them that seem to code nothing recognizable. This DNA has been labeled *junk DNA* because scientists do not yet know the purpose it serves. To compound the difficulty of finding and identifying genes, there are genes called jumping genes(jumping DNA) or *transposons* that move along that DNA or even between the DNA of different chromosomes. The entire human genome list is not tackled as a whole. Instead, scientific study is focused on small segments at a time by trying to either find genes and determine their functions or look for abnormalities in genes responsible for disease. Here, comparing genes from different genomes listed in a genome bank can be helpful. But one complication is that researchers now realize that not all diseases are monogenetic, i.e. one gene causes one disease. In fact, a very small percentage of diseases, perhaps 2%, are linked to only one gene, as is the case with muscular dystrophy. Most diseases are multi-

guish the chromosomes from one another. There are 24 different chromosomes for humans. 22 are called autosomes and two are the sex chromosomes X and Y. Each human cell (except the reproductive cells) contain 23 pairs of chromosomes in the nucleus of each cell—one set of 23 from each parent. Chromosomes have anywhere from 50 to 250 million bases of DNA, and in addition, contain proteins and RNA. The specific atomic formations which allow only A to pair with T and C only with G. These combinations—AT, TA, CG, and GC— resemble rungs of a ladder with the ladder's sides composed of a sugar and phosphate. Because of the pentagonal shape DNA sugar, the ladder makes a complete twist after every ten rungs, and in this way it forms into DNA's *double helix* shape. Among the bases within the DNA are *genes*. Genes are referred to as units of heredity. The bases composing a gene are actually instructions for producing proteins, replicating DNA, and carrying on necessary functions of a cell. *Each gene itself is a particular program.* The thousands of genes along the DNA molecule are analogous to the "on" and "off" switches of a computer's binary instructions, but these are genetic programs or algorithms. Genes are triggered into action by various stimuli, such as exposure to a particular hormone. A stimulated gene uses a chemical reaction to trip the switches of other genes. Its signal runs along the DNA turning genes "on" and "off" Those turned off stop transmitting a signal while those turned on begin transmitting their own signals. Thus the gene's program may indicate when to switch other genes on or off, when to make certain proteins, etc.

Whose DNA was used for the HGP? An anonymous individual's DNA has been sequenced. And other individuals' DNA of diverse ethnicity are also being catalogued. The DNA sequence of individuals differs very slightly. It's these subtle changes in the order of the DNA bases that account for each of our unique characteristics. It is startling to consider a human's DNA has only 300 unique genes that set it apart from a mouse's.

Like nucleic acids, *proteins* are also long chains of smaller units called amino acids (thus far 20 types of amino acids have been identified). Every protein is identified by a specific sequence and number of amino acids. *Enzyme proteins* are responsible for directing the chemical reaction of a living cell. *Where does RNA enter the picture?* When the DNA is splitting during its replication process, a messenger RNA is constructed along the DNA strand matching complement bases on the DNA strand. (This process is called transcription.) The messenger RNA carries this genetic code from the DNA to the translators of the genetic code called transfer RNA molecules. Here a string of amino acids are connected in the translated sequence by ribosomal RNA and the protein is formed. Many steps and parts have been eliminated for simplicity, but the wonder of this process taking place simultaneously in living cells is mind boggling. Proteins and *enzymes* (proteins that do not change after doing their job) perform the functions that sustain a cell's life.

factorial —"are influenced by many genes interacting with one another and by a vast array of signals within the cellular environment (including nutrient supply, hormones, and electrical signals from other cells), and these in turn are influenced by the external world or the organism as a whole."[3]

How are scientists tackling and unraveling the secrets within the human genome? With *bioinformatics*, a field that connects biology, computer science, and mathematics. Biologists are refining techniques for separating, studying, and reconnecting fragments of DNA using genetic engineering methods which include replicating, altering, splicing and attaching/reattaching code sequences. *Codons*, which are strings of three of genetic bases, tell which amino acids are used to make up specific proteins. Each protein starts with the codon ATG and ends with one of three other codons (TAG, TGA, or TAA). But, just looking for these stop and go signs to pinpoint proteins is not as easy as it might seem. Even though the base codes for amino acids and codons are known, what comes before or after them may influence how they are identified. For example, ...CGAAGAC...

could be read as ... CGA AGA C... or ...C GAA GAC... or... CGA A GAC... where CGA is the codon arginine, AGA is arginine, GAA is glutamic, and GAC is aspartic acid, etc. Thus, the way a sequence of bases is divided up in translating it makes a difference in which codon is identified. In addition, each protein has a constellation of amino acids called its *proteome*. The proteome's constellation is not fixed but in a state of dynamic flux. These dynamic changes occur in response to thousands and thousands of environmental factors taking place inside and outside the cell while the gene's sequence directs a protein's activities.

Nearly half of the human genome is composed of *transposons* or jumping DNA. This was first recognized in studies by Dr. Barbara McClintock in the 1940s. In the 1980s scientists confirmed her findings by observing transposons in other genomes. In 1983, McClintock received a Nobel prize for her work. Today scientists believe these transposons were important players in human evolution and play important roles in some genetic disorders, such as hemophilia and leukemia.

Computer programs have been developed and are constantly being improved to sort through all the possible combinations. Researchers are looking for patterns which might indicate parts of DNA that are genes, protein-coding regions, RNA, etc.

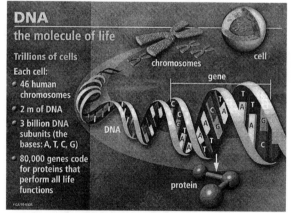

Courtesy of U.S. Dept. of Energy *Human Genome Program*
www.ornl.gov/hgmis

But sorting is just one challenge. Deciding which is the actual intended match is very difficult. Here complex statistical analysis is used to arrive at the best possibility of how the bases should be translated in the context of what comes before and after them. *Gene hunting programs* and *protein forming programs* have also been developed. What is frustrating is that different programs give different results, thereby requiring additional analyses and comparisons to guess the best possible interpretation. Programs have been devised to compare short sequences of DNA fragments to known sequences in a DNA bank, where the genome sequences of over 60,000 species of animal, plants, bacteria, viruses are stored. All organisms' DNA — human or not — is composed of the same four bases. This area of study is referred to as *comparative genomics*. But, "comparative genomics doesn't make distinction between genes and junk(DNA)...whole comparisons are the only good way to look for regulatory regions of genes"[4] making DNA banks essential to the process of mapping the HGP.

The key to all this seemingly endless search lies in identifying and recognizing patterns. The process of finding patterns in the sequence of the bases of a genome is called an *annotation*. Even as some books are annotated to clarify and expand what is being said, similarly the genome is annotated when a gene or significant pattern of bases is identified. The list of bases of an unannotated genome is essentially useless because it is the annotations that unravel its secrets. But even finding annotations doesn't necessarily tell the scientists how the cell utilizes this information. The next step is figuring out the function of that pattern for the cell.

Scientists have not been able to access and script every base in the human genome, and it is estimated that the completed human genome will have 90% of the genes mapped. How can we get the whole picture with parts missing? Mathematics will be an invaluable player. Cryptographic techniques are being used to uncover and unlock patterns and functions hidden in DNA codes. Computer modeling is being used to examine the possible scenarios and uncover annotations for the sequence of

DNA bases. Knot theory studies the possible significance of the configurations found in DNA molecules. Chaos theory can be used to study how the subtle changes of bases may produce profound outcomes in evolutionary process. Probability and statistics are used to try to pick out the likeliest translations and interpretation of the bases making up a DNA fragment. And computer programming, genetic programming and algorithms are the tools for analyzing and dealing with the volumes of raw data from this formidable project. "Scientists will be making major discoveries from the human genetic code a hundred years from now."[5] Will these future discoveries   lead to a new revised

Knowing the 3-D structure of a protein, which acts as a switch governing the cell growth, may enable intervention to shut down this switch in cancer cells.
Courtesy of U.S. Dept. of Energy **Human Genome Program** *www.ornl.gov/hgmis.*

theory of genetics? Will the current *theory of genetic-determinism (the theory that genes basically determine evolution) be overshad*owed by the *theory of dynamic-epigenetics* (the theory that complex dynamics and interactions taking place both inside the cell and in the outside environment drive the evolution of life)? Only time will tell.

[1]The Monk in the Garden by Robin Marantz Henig, Houghton Mifflin Co., Boston, 2000.

[2]A New Paradigm for life by Richard Strohman, California Monthly, April 2001.

[3] A New Paradigm for life by Richard Strohman, California Monthly, April 2001.

[4]The Meaning of Life by Tina Hesman, Science News, April 29, 2000.

[5] Quote by J. Craig Venter, president of Celera Genomics, from The Meaning of Life by Tina Hesman, Science News, April 29, 2000.

the mathematics
of peace

S ome phenomena in life seem predictable. For example, the
Earth rotates every 24 hours on its axis as mathematical
equations and physical laws predict. Yet, so many aspects of
our lives are better described by complex systems which
sometimes appear orderly and predictable and other times
totally chaotic. Chaos is the state in which existing natural laws
cease to exist, work, describe, or predict natural phenomena.
Complexity theory is an area of mathematics devised to study
systems which exhibit both order and chaos. The peace process

2 suspected bombers
killed by Israeli tank
Work begins on electronic fence
to close off 3 sides of Jerusalem

U.S. delays
veto of
U.N. troops
in Bosnia
Security Council
allowed 12 days
to solve court crisis

Muslim party
in Kashmir
softens policy

Koreans
trade
blame
after
skirmish

Mexican water
problem resolved

Israel to allow
Palestinians
out during day

India leader names
hawk as his successor

U.S. ignores abuse
of Pakistani women

Peru's war on coca
halted after protests

Pakistanis want terrorists,
Americans out of country

Irish police quell
rioters with water

Russians doubt Chechens
ready to take on rebels
Putin wants troops out, but officers dubious

U.S. threatens
to shut down
Bosnia mission

also takes place in such frameworks. Events taking place everyday on the Earth, regardless of how globally insignificant or imperceptible, can eventually affect the delicate balance between order and chaos, between peace and war. A nomad's horse is stolen in the Sahara desert, a farmer's crop

accidentally destroyed by skirmishes between two feuding factions, or a terrorist group in Ireland, Palestine, USA or Israel launches a surprise attack on innocent citizens. Each of these acts can eventually influence the delicate balance of peace, tittering back and forth between order and chaos. Such a system is perpetually in constant flux adjusting and readjusting between chaotic and ordered states.

Today mathematicians and scientists are considering the properties that comprise complex systems in the hope of unveiling some sort of predictability. They seek to identify patterns and other conditions that might indicate the emergence of a chaotic state. It seems almost impossible to think that chaos can be mathematized, especially since its name is the antithesis of order and predictability. A major breakthrough in chaos theory first surfaced in the late 1970s with the work of physicist Mitchell Feigenbaum. While comparing numbers generated from a quadratic iterative equation, he discovered that a certain constant ratio continually reappeared — $4.6692... = \delta$. This constant has turned out to be very important and influential in our lives. Regardless of which the iterative equations Feigenbaum tested, the same number ratio always appeared. He felt so strongly about his discovery that he conjectured that it was a universal constant for chaos theory. His work gained credibility in 1979

> *Complexity theory deals with complex systems which exhibit both order and chaos, not necessarily at odds with one another, but paired in a dance moving back and forth from the edge of chaos.*

In a **complex system** the same set of circumstances do not necessarily produce the same results, but minute/ imperceptible changes influence these nonlinear systems.

when computer scientists Oscar Lanford III, Pierre Collet and Jean-Pierre Eckmann devised a computer assisted proof of the Feigenbaum conjectures. This work set in motion many more mathematical studies into chaos theory. The Feigenbaum discovery and Feigenbaum constant show at what point order can revert to chaos. In other words, these numbers can help predict when a system will become chaotic.

How can an iterative process be connected to peace? Mathematically speaking an *iterative equation* continually generates new results by recycling its last answer. This is done by plugging it back into the equation so a new result is spewed out. This process is repeated continually, in other words reiterated. A beginning value is run through an equation; for example, consider the equation: $x(x+1)=new\ value$. An initial value for x is arbitrarily given, suppose for this equation it is x=5. Then, the following numbers are generated by $x(x+1)=new\ value$: 30, 930, 865830,... . These numbers don't seem to possess any interest. Yet it was in such an equation that Feigenbaum made his discovery.

Now consider a natural iterative process, population growth. We begin with a set population, which produces offspring, which in turn are used to produce offspring... the process resembles an iterative equation, except that many factors surface that influence the process and the resulting population. For example, suppose not all offspring survive, suppose only one sex of offspring is

$$x_{new} = 3x_{initial} + C$$

$$\text{let } C = 2 \text{ \& } x_{initial} = 1$$

the values generated are

5, 17, 53, 161, 485, ...

produced, suppose there is a food shortage, an outbreak of disease,... the possible scenarios are endless. Unlike the mathematical iteration above with its single starting value and arbitrary constant C, population continually evolves and changes, and is influenced by innumerable conditions with elements of random behavior surfacing.

Consider peace. It is a process, rather than a perpetual final state. A process whose balance is ever so delicate. So many factors — some major, such as peace talks — some globally imperceptible, such as food shortages in remote villages in the world — food and medical supplies distribution to global disaster areas — a pleasant greeting to a stranger — all influence the balance between peace and war. No matter how small or seemingly insignificant a positive or negative act may appear, its effect is no less important than of the fluttering of a butterfly on global climate. The mathematics of chaos and the Feigenbaum constant, along with complexity, statistics, probability, fractals geometry, and fuzzy logic perhaps will provide the necessary tools for analyzing global phenomena and their profound effects.

the millennium clock

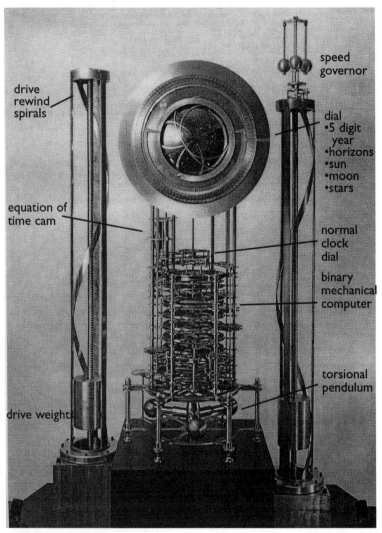

drive rewind spirals

speed governor

dial
•5 digit year
•horizons
•sun
•moon
•stars

equation of time cam

normal clock dial

binary mechanical computer

torsional pendulum

drive weights

Photograph courtesy of the Long Now Foundation

To keep track of **time,** humans have invented a host of devices. They've marked the seasons with Stonehenge, the hours with sundials and waterclocks, the seconds with pendulums. We are obsessed with measuring movement and speeds. How many seconds does it take your car to go from 0 to 60mph? At how many bauds does your modem run? How about an atomic fountain clock, so accurate it marks nanoseconds (1-billionth of a second)? And of course, we have our own personal body clock that measures one's lifetime. What about slowing things down by designing a clock that ticks once every year, and chimes every millennium? Such a clock was first envisioned by computer scientist Daniel Hillis in the early 1990s. Hillis and other members of the Long Now Foundation made the first prototype, which is now housed in the London Science Museum. The millennium clock is a combination of high and low tech ideas. It could have been designed to be an electronically self-driven clock, but instead it was purposely designed to be manually wound, thereby requiring humans to be responsible for its functioning. Why? To make us focus on the larger picture of the universe, and help us realize that what we do now is for the future.

What does it mean to say the time is accurate? As of 1999, the globally accepted definition of a unit of time was the period when 9,192,631,770 oscillations of a certain microwave frequency emitted by the cesium-133 atom takes place. A precise clock is defined as one whose frequency variations from this is minimal.

The Long Now Foundation was established in 1996 and "seeks to promote 'slower/ better' thinking and to focus our collective creativity on the next 10,000 years." (http://www.long now.org)

The plan is to eventually have a millennium clock installed near the summit of a mountain, which will need to be wound manually once a year. Hillis wanted to emphasize the need for human participation with the passage of time. Why a 10,000 year design clock design? Hillis explains that our species has been around for the first 10 millennia, and now we look to the next 10,000 years. In addition, the millennium clock will help us look

"Time has no divisions to mark its passage. There is never a thunderstorm or blare of trumpets to announce the beginning of a new month or year. Even when a new century begins it is only we mortals who ring bells and fire pistols."

—Thomas Mann

beyond ourselves, and our daily needs — our 30 year mortgages, our immediate stock gains — to what future we sow by our actions. As Hillis says, "I think of the oak beams in the ceiling of College Hall at New College, Oxford. Last century, when the beams needed replacing, carpenters used oak trees that had been planted in 1386 when the dining hall was first built. The 14th century builder had planted the trees in anticipation of the time, hundreds of years in the future, when the beams would need replacing....I cannot imagine the future, but I care about it. I know I am part of a story that starts long before I can remember and continues long beyond when anyone will remember me. I sense that I am alive at a time of important change, and I feel a responsibility to make sure that the change comes out well. I plant my acorns knowing that I will never live to harvest the oaks."[1]

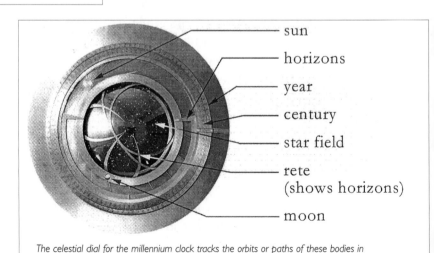

sun

horizons

year

century

star field

rete
(shows horizons)

moon

The celestial dial for the millennium clock tracks the orbits or paths of these bodies in relation to the year. Photograph courtesy of the Long Now Foundation.

Utilizing his computer skills, Hillis designed the clock with a binary digital mechanical system with precision equal to one day in 20,000 years. In addition, it self-corrects by phase locking on celestial bodies. The clock keeps track of the

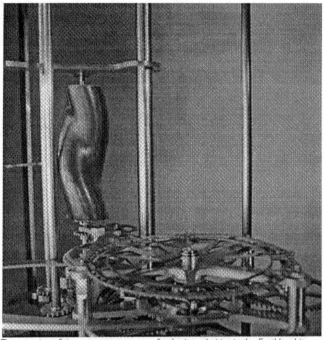

The equation of time cam compensates for the irregularities in the Earth's orbit over time. The cam is fashioned as a twisted cylinder designed to measure the 26,000 year precession of equinoxes. Photograph courtesy of The Long Foundation.

rising and setting times of the moon and sun, it follows the phase and position of the moon, its "bit adder"[2] digitally measures time and programs into the clock the unique characteristics of the Gregorian calendar, the equinoxes and the solstices for 10,000 years.

The millennium clock is evolving in prototype stages. The clock currently on display in the London Museum of Science is 8 feet tall. It was put into operation on December 31, 1999, just in time for a millennium event. The second prototype will be about 18 feet tall, and is currently being designed/built in San Rafael, CA. What physical features do these prototypes have? (a) they are made of a nickel-copper alloy to minimize expansion/contraction with temperature changes, (b) their pendulum consists of three 22 pound tungsten balls powered by a two helical spiral weights drives which are positioned on opposite sides of the clock. (c) the equation of time cam[3] lets the clock compensate for the irregularities in the Earth's orbit over time. The cam is fashioned as a twisted cylinder designed to measure the 26,000 year precession of equinoxes.

The final millennium clock is projected to be about 40 feet in height. Whether its final form will be realized is not as important as whether people are discussing the ramification of how our actions are influenced by time, its increments, and our impact on future.

The torsional pendulum keeps time by rotating rather than by swinging as a conventional pendulum. Photograph courtesy of the Long Now Foundation.

[1]http://www.longnow.org/10kclock/clkPurpose.html

[2]"Danny Hillis' Serial Bit Adder is an all mechanical, yet digital binary calculator. It is programmable with what we call bit pins (metal dowel pins) which are set in either a 0 or 1 position adding up to the number we need to calculate. This allows the majority of all our calculations to be done without the sliding friction of gears and allows us to change the numbers later on."—Alexander Rose , designer with the Long Now Foundation.

[3]"The clock's star map in the center precesses with the equinoxes every ~26000 years and the equation of time cam also keeps track of this for our solar to absolute time correction."—Alexander Rose, designer with the Long Now Foundation.

chaos theory

Even though chaos theory is one of the cutting edge areas of mathematics, its story dates back to the turn of the century. In 1846 Uranus' orbit deviation indicated the possible existence of another planet, Neptune. Did this discovery imply Isaac Newton was correct in stating that phenomena of the universe were predictable? This question prompted the King of Norway in 1889 to issue a challenge and prize to the first person to prove whether our solar system was indeed stable. Mathematician Henri Poincaré's work was initially the winner, but there was one slight hitch — an error was uncovered in his calculations. Poincaré was given six months to correct this error, and in the course of that time Poincaré came to realize that the stability of our solar system was not predictable. His findings contradicted the current thinking of a linearly predictable Newtonian universe. In fact, Poincaré was the first to write about how small changes in initial conditions could have profound effects on outcomes. But during Poincaré's time, scientists seemed reticent to give up the comfort and

chaos theory a branch of mathematics which studies non-linear systems. Chaos is a state in which nature does not conform to established scientific laws. Before a system becomes chaotic, i.e. unpredictable, there is a transition point in which errors have been mounting that will eventually render the system chaotic, unless changes take place that will swing the system back into before the transition point was passed. Chaos crosses all disciplines and sciences.

Iterative process can involve numeric or non-numeric algorithms which begin with given initial value(s) or object(s). The algorithm is applied to the initial value or the object, and a new value or altered object results. The same process is applied over and over again to each new resulting value or object. This process can be done continually infinitely many times resulting in infinitely many values.

power of being able to "predict" outcomes by using the linear models which Newton's system provided. Although there were numerous stepping stones[1] leading to the evolution of Chaos theory, Poincaré was the first to recognize concepts of chaos in natural phenomena.

The turn of the 19th century was a time when the work of mathematicians and scientists dealing with chaotic systems was essentially ignored. Very few scientists wanted to zero in on a system's glitches, when the system worked predictably most of the time. In fact, it was not until 1975 that the term chaos was first used. Tien-Yien Li and James A. Yorke first applied the word chaos to phenomena introduced by non-linear systems in their article *Period Three Implies Chaos* in the *American Mathematical Monthly.* (vol. 82). Yet, chaos theory (also referred to as nonlinear science) did not really stir the scientific world's interest until physicist Mitchell Feigenbaum uncovered a universality that existed within chaos theory. Feigenbaum's work yielded the Feigenbaum constant, which is ratio showing at what point order can turn to chaos. In other words, the number can help predict when a system will become chaotic. This ratio kept appearing in the many diverse systems and iterative equations he tested. It was even discovered in the Mandelbrot set. But it was not until 1979 that computer scientist Oscar Lanford III put chaos theory on firm mathematical footing with his computer assisted proof of the Feigenbaum conjectures and with the work of Pierre Collet and Jean-Pierre Eckmann.

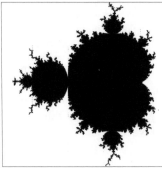

This is one of the famous fractals discovered by Benoit Mandelbrot in the 1970s while carrying out his ground breaking work in fractal geometry. He generated it using the iterative equation, Z^2+C, involving complex numbers. Starting with two complex numbers, one for the variable Z and one for the fixed constant C, the equation generates a value. That value is then replaced for Z and a new result is generated. If the process is continued indefinitely and the results graphed, a Mandelbrot set begins growing and branching in all sorts of fascinating ways. It was both startling and amazing to find that Feigenbaum's universal chaos constant was also present in this fractal.

A *dynamic system* — one that fluctuates — can become chaotic. Due to imperceptible or unmeasured changes in the initial components of such a system, an unpredictable situation can escalate sending a system into a state of chaotic (unpredictable) behavior[2]. Yet within chaos a self organizing component is also present which can eventually restore order. Along with the term chaos other terms have been adopted to describe the phenomena taking place in the world of chaos theory. Here we have self-maintaining and self-organizing dynamics, complexity, equilibrium, non-linear systems, sensitive dependence on initial conditions, strange attractors, and the well known butterfly effect.

The butterfly effect and chaos theory were first linked in 1962 when meteorologist Edward Lorenz published his findings on how minute changes in initial weather conditions can reek havoc on long range weather conditions. His famous Lorenz attractor and strange attractors in general are now universally recognized and connected to chaos theory. Chaos theory and the butterfly effect are not confined to meteorology — its influences have been noted in economics,

Self-organizing or *self-maintaining* dynamics are the mechanisms by which a system regains its equilibrium or balance, by continually changing and adapting itself to constantly changing factors or circumstances.

Linear system is a predictable system. In a linear system a particular value always gives the same result. The term linear in mathematics comes from the word line, which means the line's equation, written with the variables x and y, remains in a particular constant proportion.

Non-linear system is unpredictable. For such a system a specific value does not always yield the same result. For example, weather is a non-linear system. The weather for a particular time and place may be sunny one year and have snow on another year.

Complexity or a *complex system* is a dynamic system tittering between order and chaos. It is always in a state of flux, existing on the edge or verge of chaos where there is a continual tug of war between order and chaos.

Strange attractor is a graph of a particular system (for example, Lorenz attractor) which is graphed in multi-dimensional space. It constantly changes, endlessly looping but never exactly repeating itself. It is startling to see a chaotic system producing this fascinating "pattern" which possesses chaos and order simultaneously.

The *butterfly effect* uses the analogy of how the air movement created by the wings of a butterfly can eventually result in a full scale storm in another part of the globe. This effect illustrates that minute almost imperceptible changes can develop into enormous consequences, dubbed *sensitive dependence on initial conditions*.

ecology, cinematography, in fact in all sciences and areas.

During the 2000 US presidential election, the butterfly effect took on an additional meaning with the butterfly ballot in Palm Beach County, Florida. Statistical analyses, studies and experiments were done to try to determine if indeed the outcome was affected by the butterfly ballots. Most studies found that votes were skewed by the physical nature of the ballot. But, all that is history.

The butterfly ballot was so named because it resembled the wings of a butterfly. But above and beyond that we saw that minute changes, i.e. in a nominal number of votes, had a profound effect on the election's outcome. A fascinating coincidence. Chaos theory at work?

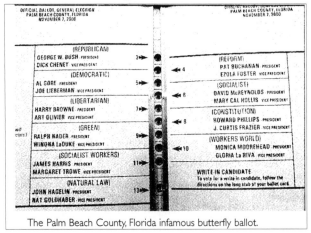

The Palm Beach County, Florida infamous butterfly ballot.

Today chaos theory is being embraced by most scientists in all disciplines. Thousands of papers have been written, courses are taught, and journals are generated about its various components and interconnections. Elements of chaos theory surface in all sciences from aeronomy to zoology, in the workings of our bodily functions to the flow of traffic on the freeways, in the control of our energy grids to electric outages to internet connectivity. Everywhere we look we see signs of chaos theory. Adam Smith's analogy of an invisible hand guiding the supply and demand of a capitalistic economy

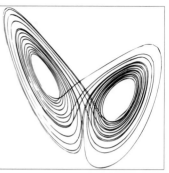

The exciting thing about the Lorenz strange attractor is that it revealed an uncanny order hidden in a chaotic system. In the early 1960s, Edward Lorenz, using a 3-dimensional space, graphed the data from his weather model on a computer programmed to register various changes in the rising of hot air. Regardless of how small the change in initial conditions he used in his equations, the results differed widely, which led to his discovery of the butterfly effect. In the 3-D space, the Lorenz attractor never intersects itself, endlessly and randomly loops and as yet never repeats its path. It can reverse at whim—a totally unpredictable chaotic path that incredibly retains an orderly shape.

is actually chaotic and complex systems at work. Here again imperceptible minute changes add up to huge economic consequences. Everywhere in the universe complex systems and components of chaos theory are at work. No scientist can ignore its existence, its influences and its impact on all aspects of our lives. Whether you're a biologist, an economist, a diplomat, or a farmer you need to know how the balance of complex systems evolves. You must be able to recognize what indicators tip the scales at the edge of chaos, how complex systems can regain their balance and self organize. Today, from seemingly simple phenomena, such as a dripping faucet to the complex functioning of the human body, chaos theory is being used to explain, describe, and predict.

---

[1]Other contributors of initial ideas to chaos theory include: Jacques Hadamard and his theorem involving initial conditions, Pierre Duhen and his 1906 paper pointing out Hadamard's theorem dealing with influences of initial conditions, and early mathematicians exploring fractals.

[2]Examples: — a small family fight in one village leads to a global war — changes in the flow of a river lead to flooding — day-trading by a single person sets the market off — minute changes in supply and demand of electricity lead an economy to a recession. All phenomena, no matter how stable they might appear, may exhibit chaos at some point — even in our body's functions, the fashion world, a dripping faucet, the weather, forest fires, the spread of a virus…politics and voting.

cricket math
— chirps & temperature

Who would have thought that the weather's temperature and a cricket's chirps are related. Just consider either of the equations: t = (c/4) +40 or turned around c=4t - 160.

These equations relate Fahrenheit temperature, *t*, to the number of chirps per minute, *c*, of a cricket. Have you ever heard any cricket chirps during really cold weather? If the temperature is freezing you won't hear any. In fact, if the temperature is 40° the equation *c = 4t–160* shows there are 0 chirps. But at 45°, the chirps are at 20 per minute; c = 4(45) -160= 20.

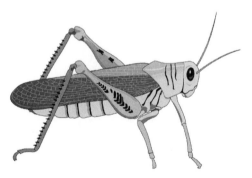

On the other hand, the chirps become louder and faster the moment the temperature reaches the high 70's. At 80°, the number of chirps is 160 per minute is

c = 4(80) -160=160.

Or check out the temperature when the chirps get up to 200 per minute,

t = (200/4)+40 =90°.

This natural thermometer won't replace our high tech ones, but it's fascinating to discover it.

mathematical private eyes

ratios hold the clues

Yes, old fashioned fractions are helping archeologists, art historians, geologists, and even the DEA(Drug Enforcement Administration) to uncover and analyze information. Mathematics and the science of chemistry are joining forces to produce data bases of mathematical ratios. These fractions, called isotropic ratios, compare quantities of various elements present in different types of matter. Isotropic ratios are being used to discover ancient export/import routes, to identify counterfeit sculpture, and even to help the DEA pin point coca growing regions from samples of confiscated cocaine.

Consider marble, for example. Mount Pentelikon in Greece has over 150 quarries along its slopes. Over the past 2500 years, these quarries were the primary supply sites of white marble for the Eastern Mediterranean region. Until the isotropic ratio method was developed, archeologists and historians relied on a stone's texture and grain in order to determine its origin. A data base of carbon and oxygen isotropic ratios for this region was compiled by analyzing over 600 samples collected from the quarries of Mount Pentelikon. A specific piece of marble's isotropic finger-print is given by the ratio of carbon & oxygen atoms present in it. These ratios give scientists the necessary information to identify from which quarry a specific piece of marble came. Such information

> *ratio:* The relative size of two numbers or quantities represented by the quotient of one divided by another. The ratio of a to b can be written as a:b or expressed as the fraction a/b.

Columns at the oracle of Delphi

will help art historians discover clues about where a sculpture was done, and also help determine the possible sculptor of a particular piece. In addition, the isotropic ratios for a hunk of marble are very consistent making it feasible to identify pieces of work carved from the same piece of marble. Further, isotropic ratios would help art historians determine forgeries.

The value of isotropic ratio fingerprints is not confined to such long lasting matter as marble, but is also very useful with the nondurable matter of plants. The chemical fingerprint of a coca leaf can point to the region of its origin. The isotropic ratios for coca leaves are determined by comparing the ratio of its carbon and nitrogen atom. The various ratios of coca plants arise from the unique conditions of the region in which they were grown. These conditions include soil content and climatic factors. Until this isotropic ratio analysis method was developed, the DEA had to rely on determining the origin of cocaine by trying to analyze the chemicals used to process it. The isotropic ratios help identify where the leaves originated, rather than where they were processed thereby helping the DEA identify drug traffic routes, cultivation areas, and regions to pinpoint for eradication.

Who would have thought that fractions would be used on one hand to identify the origin of an ancient piece of sculpture and on the other play have a major role in a drug bust?

Coca is a densely-leafed plant native to the eastern slopes of the Andes.

holyhedrons

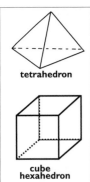

tetrahedron

**cube
hexahedron**

octahedron

dodecahedron

icosahedron

Polyhedra have been around since ancient times. The tetrahedron, the cube and the dodecahedron date back to the Pythagoreans. In fact, the five Platonic solids were first written about in Euclid's book, *The Elements* (circa 300 B.C.). It was Plato's friend, Theaetetus, who proved that only five regular convex polyhedra exist — the tetrahedron, the octahedron, the hexahedron(cube), the dodecahedron, and the icosahedron. These are the only convex solids whose faces were composed of congruent convex polygons.

But leave it to a mathematician to look at a polyhedron with a new twist, and come up with the holyhedron — a polyhedron with a hole passing through each of its faces that still remains a polyhedron. What polyhedron can be holed in some way so that resulting figure is a polyhedron? Mathematicians thrive on solving such problems.

Polyhedra are solids whose faces are in the shape of polygons. A polygon is *regular* if its sides and angles are congruent. A polyhedron is regular if all its faces are identical, all its edges are the same size, and its corners have angles the same measure.

A polygon is *convex* if all the points between any two points for any face of polygon are also on that face of the polygon.

The problem originated several years ago, when mathematician John H. Conway of Princeton University was asked if a polyhedron with a hole on every face could exist. He precisely defined the problem and put up a reward for its solution — $10,000 divided by the number of faces of the holyhedron solution , and put it out there for mathematicians to stew about. As part of his definition, Conway required that the holyhedron be finite. Imagine producing the holes by piercing each face of the polyhedron with a vertex of another polyhedron. This process introduces new faces that also have to be pierced, which produced a solution that required an infinite faced holyhedron.

A rendition of how part of a lattice of an infinite number of tetrahedra would pierce one another to form an infinite "holyhedron".

Working along these lines, mathematician Jade P. Vinson came up with a humongous holyhedron which had 78,585,627 faces. After dividing the $10,000 by the number of faces of this holyhedron Vinson's reward equaled about $1.27. Money was obviously not Vinson's motivation. His method involved joining a large finite number of convex polyhedra that were arranged so that every face of every polyhedron was pierced by the vertex of another. He determined that the set of convex polyhedra he used had to have more vertices than faces. Vinson is confident that there are holyhedron with fewer faces. Other mathematicians have built and expanded upon his work , or taken other avenues. As with all interesting problems, one thing leads to another. New related questions and problems have popped up, such as— *What is the minimum number of faces that a holyhedron can have?*

Can a physical 3-dimensional model of a holyhedron be constructed? What will all this holyhedron stuff lead to? There may be some surprises in the future.

fractals, fractals everywhere

Before 1960 the term fractal didn't exist. Before then the objects now referred to as fractals were considered mathematical anomalies. When they first began appearing in the works of 19th century mathematicians, they were generally referred to as monsters. To most mathematicians of this time, these strange shapes were thought to be counter intuitive. Fractals contradicted accepted mathematics and seemed almost paradoxical — some were continuous functions while not differentiable, some had finite areas and finite perimeters, and some could completely fill space. *So what happened in the 1960s?*

From the 1860's to the early 20th century mathematicians who explored fractals included: Georg Cantor, Helge von Koch, Karl Weierstrass, Dubois Reymond, Guiseppe Peano, Waclaw Sierpinski, Felix Haussdorff, A.S. Besicovitch (Haussdorff and Besicovitch worked on fractional dimensions), Gaston Julia, Pierre Fatou (Julia and Fatou worked on iteration theory), and Lewis Richardson (worked on turbulence and self-similarity).

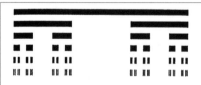

The Cantor set illustrates both the occurrence of batches of error transmissions and error free regions. The segments in the set represent the batches of errors and the open regions represent the error free gaps. In 1883 Georg Cantor's mathematical idea was considered by most as a curiosity. Now, over 80 years later, it has become an invaluable notion.

Mathematician Benoit Mandelbrot made a ground breaking discovery. He noticed a connection between these "weird" mathematical objects and a natural phenomenon. While working at Bell laboratories, he found that the distribution of telephone errors roughly resembled the Cantor set of 1883. He noticed errors occurred in batches with error free regions in between, and he remembered this pattern in the elements of the Cantor set. When he zeroed in on a batch, he noticed the smaller part he was examining resembled the large batch. In other words, he even noticed self-similarity in the error distribution. In Mandelbrot's book, *The Fractal Geometry of Nature*, he points out *"We construct the set of errors by starting with a straight line, namely the time axis, then cutting out shorter and shorter error free gaps. This procedure may be unfamiliar in natural science, but pure mathematics has used it at least since Cantor. As the analysis is made three times more accurate, it reveals that the original burst is intermittent."* His discovery illustrated that fractals could be used to describe natural phenomena. From then on fractals and nature have been an inseparable duo. Ever since that revelation, thousands of articles, papers, and books about fractals have been written, and fractals have been the topic of the thousands of discussions, lectures and classes.

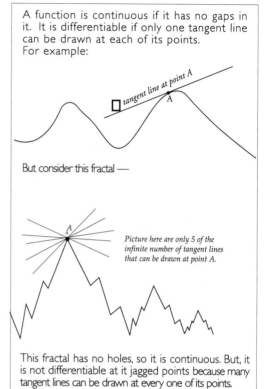

A function is continuous if it has no gaps in it. It is differentiable if only one tangent line can be drawn at each of its points. For example:

tangent line at point A

A

But consider this fractal —

A

Picture here are only 5 of the infinite number of tangent lines that can be drawn at point A.

This fractal has no holes, so it is continuous. But, it is not differentiable at it jagged points because many tangent lines can be drawn at every one of its points.

Why are fractals everywhere? Fractals are very different from the mathematical figures we learned in Euclidean geometry. Fractals have imbedded in them a type of mathematical "genetic" code that makes them dynamic figures. As such they are able to describe dynamic things, things that change and things that are alive, in addition to any shape in nature. Once conceived, a fractal seems to have a life of its own. Euclidean objects, such as perfect triangles, squares, circles do not actually exist in our

universe, but merely represent imperfect squares and circles, etc. of our world. Fractals, however, can be adapted for the odd and varied shapes and forms, and most importantly, forms that are not static. A host of musicians, artists, doctors, economists, meteorologists, astronomers, chemists, biolo-

This driftwood is one of the many odd shapes fractals can describe.

gists, physicists, botanists, computer scientists, communication specialists, and mathematicians have worked with fractals connecting facets of their work to fractal geometry. Today we even find fractals being used to both help diagnose and find a cure for osteoporosis, to determine the health of a heart, to design cell phone antennas, to track the spread of contagious diseases. The rings of Saturn have been found to resemble the Cantor set. The growth of bacteria follow fractal patterns. Fractal properties appear in aspects of nature from plant formation to the formation of lung tissues and bones. We find fractal properties present in both the large and small — in an oak tree or at the atomic level of DNA. Fractals can also describe the ups and downs of the economy, population growth and changes, and even the diffusion of gases, oils and wild fires. Fractals have been used to compose music, create art, and design artistic landscape for movies.

What properties do fractals have that make them so special? — self-similarity, iteration, and fractal dimensions. *Self-similarity* means that regardless of how much you magnify a portion of a fractal, that fragment will be similar to the original form. Nature is full of examples of self-similarity. Just look at the veins of a leaf, the branching process of a maple tree, and self-similarity will be there. *Iteration* can generally be described as the continual repetition of a particular algorithm following a prescribed procedure. The iteration process is partly responsible for the continual reappearance of the pattern, and hence its self-similarity. *Fractal dimensions* are characteristics unique to fractals, unlike traditional Euclidean whole number dimensions.[1] Fractal dimensions, on the other hand, are fractions whose numbers are usually between 1 and 2 dimensions or between 2 and 3 dimensions.[2] *What does a fractional dimension mean?* A fractal with a dimension lying between 2 and 3 would appear as a very complex looking figure. The number essentially describes how much of the region the fractal can fill. The greater the jaggedness the higher the number.

What is a fractal? A fractal is made by selecting an object which may be a geometric object such as a square or a mathematical object such as a number. A rule called an algorithm or a formula is applied to the object— making a new object. Then the rule is applied again to the new object, and another object is made. The rule is applied again, and again and again. The process of reapplying the rule or algorithm is called an iteration. The result may be a geometric fractal such as the Koch snowflake that exhibits symmetry, or you may get a random fractal, such as the Mandelbrot set. Whichever type of fractal you get, it will exhibit the property called self-similarity — *whichever part of the fractal you zero in on or magnify will resemble the original part.*

Self-similarity. Notice the repetition of the similar shape throughout the Mandelbrot set.

How do you determine the dimension of a fractal? By using this formula which mathematicians have developed $D = \log N/(\log(1/r))$.

To determine the dimension, D, of a fractal, look at a portion of it and at random select a fragment that you see repeated in that portion. Count how many copies of that fragment are in that portion. The number of copies refers to N in the formula. r represents the size of the fragment to the portion you are observing.

The frequency (N) and magnitude of a fractal (in terms of r) can be plotted on a logarithmic graph. Such a relationship is called a *power law relationship* where its slope is an exponential power which is also its fractal dimen-sion. Researchers can use fractal dimensions to study a variety of problems. Among these:

• the activity and stress along fault lines are determined by calculating the fractal dimension (in other words, its power law relationship between the numbers of rocks verses their sizes) of crushed rocks in the fault

• the occurrences of forest fires by comparing their frequency and magnitude

• the spread of urban sprawl by measuring the fractal dimension of its pattern

• comparing the fractal dimension of the spread of toxic elements through soils or ground waters

• fractal dimension of volcanic ash can help determine the types of gases, the intensity of the eruption, and volcanic history

• fractal dimensions can also be used to predict earthquakes by comparing their frequencies and magnitudes and creating computer simulations from the data.

Not all fractals exhibit *exact* self-similarity, in which the repeating pattern is obvious. Many fractals, such as *nonlinear* or *random* fractals, whose shapes do not seem to have a repeating pattern or symmetry, possess what is called *statistical* self-similarity whose self-similarity is determined using statistics. Fractals with *exact* self-similarity were used below to illustrate the calculation of a fractal's dimension,

Look at this portion of the Koch snowflake fractal. Consider this fragment. It appears 4 times in this portion, so $N=4$. It is also $1/3$ the size of the entire portion, so $r=1/3$. So the dimension of the Koch snowflake fractal is: $D = \log 4/\log 1/(1/3) = \log 4/\log 3 = 1.261\ldots$

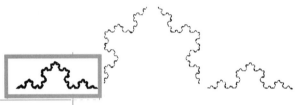

Look at this portion of the Sierpenski triangle gasket. Consider the shaded fragment. It appears 9 times in this portion, so $N=9$. It is also $1/4$ the size of the entire portion, so $r=1/4$. So the dimension of the gasket is:
$D = \log 9/\log 1/(1/4)$
$= \log 9/\log 4$
$= 1.5849\ldots$

For Menger sponge, the number of copies of the shaded fragments that appear is 20, and they are $1/3$ the size of the original. Its dimension is:
$D = \log 20 / \log 3$
$\approx 2.7268\ldots\ldots$

Mandelbrot discovered fractal dimensions by noticing that the ratio between a fractal's consecutive generated forms is constant, and this constant is referred to as that fractal's dimension. He mathematically worked out a formula to determine a fractal's dimension — $D = \log N / (\log(1/r))$. The greater the dimension number, the more complex the fractal's curve.

How are these fractals made? Fractals can be created in many ways — using mathematical expressions, computer programs, and even paper folding. Behind the idea of a fractal lies the concept of infinity with all its anomalies, paradoxes and problems. In the 19th century, when fractals were first discovered, their forms were somewhat primitive compared to today's computer generated fractals. Mathematicians exploring them could only illustrate them with pen and paper. It was at this time that the Weierstrass function, the Koch snowflake, the Cantor set, and the Peano curve were created.

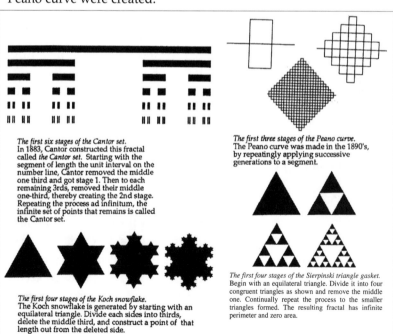

The first six stages of the Cantor set.
In 1883, Cantor constructed this fractal called *the Cantor set*. Starting with the segment of length the unit interval on the number line, Cantor removed the middle one third and got stage 1. Then to each remaining 3rds, removed their middle one-third, thereby creating the 2nd stage. Repeating the process ad infinitum, the infinite set of points that remains is called the Cantor set.

The first three stages of the Peano curve.
The Peano curve was made in the 1890's, by repeatedly applying successive generations to a segment.

The first four stages of the Koch snowflake.
The Koch snowflake is generated by starting with an equilateral triangle. Divide each sides into thirds, delete the middle third, and construct a point of that length out from the deleted side.

The first four stages of the Sierpinski triangle gasket.
Begin with an equilateral triangle. Divide it into four congruent triangles as shown and remove the middle one. Continually repeat the process to the smaller triangles formed. The resulting fractal has infinite perimeter and zero area.

What gives fractals their life? Three things— an object, a rule, and the notion of infinity. These three ideas when mixed together breathe life into this amazing mathematical idea. For example, take a regular pentagon. Draw in its diagonal, and notice the pentagram formed. But don't stop here. Draw in the diagonals of the little pentagon inside the center of the pentagram. Do this process indefinitely and a fractal evolves. A fractal can grow inward and outward. To notice this, extend the sides of the original pentagon. Notice how they intersect in a larger pentagram. If the vertices of this pentagram are connected a new larger pentagon is formed. When its sides are extended, a new larger pentagram is formed. This process can also be carried out indefinitely. To create the Waclaw Sierpinski's triangle gasket (1916), again begin with a object. Here it's a blackened equilateral triangle. Like the Cantor set, instead of adding something, an equilateral triangle is removed from its center as shown on the previous page. This process is repeated with each black triangle. When these iterations are carried out infinitely the Sierpinski triangle gasket fractal appears. A fractal can be formed just by *folding* a piece of paper in a certain way an infinite number of times.

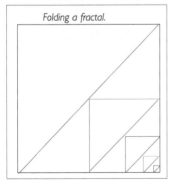

Folding a fractal.

Fractals formed using mathematical formulas are invaluable for storing data and describing complex systems. For example, instead of storing all the pixels of a scanned photo, fractal functions are stored and used to recreate it from scratch, thereby requiring far less computer memory.

The fractal craze took off in in the late 20th century. Although initially some mathematicians felt it was only a passing fad, it seems everywhere one turns fractals are popping up, and their invaluable scientific applications are being continually discovered. Just as nature and its functions are not always predictable, but often exhibit elements of chaos and dynamic non-linear systems, so fractals are being

Fractals & cell phones? Yes! Sierpinski triangle gasket or square fractals are being used to design cell phone antennas because of how well their self-similarity (the continual repetition of a particular pattern appears through a fractal) works at all frequencies.

adapted to the randomness and complexity of life. The 21st century will not only witness the interconnections and workings of the mathematics of fractals, chaos theory, fuzzy logic, and complexity, but benefit from their applications.

A forest of fractal trees. Notice how the repetition of a fractal appears in the trunk and limbs of the trees.

---

1 In Euclidean geometry we have the 0-dimensional point, the 1-dimensional segment, the 2-dimensional plane, the 3-dimensional solid.

2Mathematically speaking there are many types of dimensions. Among these we have Euclidean dimension, topological dimension, Hausdorff dimension, fractal dimension, box-counting dimension, self-similarity dimension, capacity dimension, information dimension, etc. One could almost say dimensions are related to fuzzy numbers, assuming a range of values between 1 and 2 and 2 and 3, etc.

smart dust, MEMS

What's it all about?

Ever wonder how your automobile knows when to deploy your airbag? A MEMS tells it. The MEMS sensor measures a car's deceleration. For example, when a collision occurs, the car's speed suddenly drops, and the MEMS triggers the airbag to inflate instantaneously. Before MEMS were available, the airbags were triggered by a much larger electromechanical devices about the size of a can of soda and weighing several pounds. In contrast, an airbag MEMS is about the size of a half-inch cube, making it practical to have airbags in other parts of the automobile such as in the doors for side impact. Unlike the early large units, the MEMS carries out a self-test every time the car is started to insure the system is operating. If the test fails, an indicator lights up on the dash board.

MEMS—*microelectromechanical system(s)* also referred to as micromachine sensors which has moving parts, joints, springs, gears, levels, etc. These machines are as small as fleas, ranging anywhere from between 0.1 and 100 microns.

Today MEMS are one reason your cell phones can be made smaller and with more features. MEMS are being use in areas where liquid crystals are not as effective, such as in high-definition TVs, PC projectors, and digital cinemas. There are MEMS which sense heat, movement, light, biological molecules, chemicals.

Like computer chips, MEMS are also manufactured by etching a silicon wafer. A MEMS can be made with thousands of sensors etched on a single wafer. But unlike a microchip, which uses the electrical properties of silicon, the MEMS are machines which also use the chip's mechanical properties.

Getting a perspective on the size of a mote. If this mote were to be circumscribed by a sphere, the sphere 's volume would be 11.7 cubic millimeters.
Photograph courtesy of http://www.eecs.berkeley.edu/~pister/SmartDust.

Many of us have heard of smart cards, smart washing machines, smart dryers, smart thermostats— but what about smart dust? And yes, MEMS play a role in smart dust. The idea and name of smart dust first occurred to Kris Pister in 1992. He envisioned tiny units, also called *motes,* that would consists of a miniature computer, a sensor (MEMS) , and its own wireless communication capabilities and a power source. Since computers, wireless communications devices, and MEMS were continually diminishing in size— he envisioned motes eventually becoming the size of a speck of dust. TinyOS, a computer operating system designed by David Culler, joined forces with Pister's motes and successfully demonstrated to the DARPA(Defense Advanced Research Project Agency) that tiny motes could be very effective and useful. As a test, a small remote controlled plane dropped proto-types of smart dust units over a military base. These motes were equipped with magnetic sensors. Tiny OS allowed these little motes to immediately work together. By reading magnetic signatures, the

> MEMS and smart dust are big business. As of 2001, it was estimated that there are 1.6 MEMS devices per person today in the United States. By 2004, it's projected that there will be 5 devices per person.

motes were able to detect and report what type of vehicles passed by. The possible uses and misuses of smart dust are far reaching. They can be designed to monitor traffic flow or to create a virtual keyboard or translate sign language by attaching smart dust units to your fingertips. Smart dust motes could be used in a building's sensor network which could be used in case of fires. In collapsed structures they would continue to function

detecting heat and sounds of survivors and relaying the locations. In the business world, ima-gine smart dust used in warehouses and trucks to monitor inventories from a palette of goods to the shelf life of a single item.

MEMS size: 1mmx25mmx28mm
Photograph courtesy of http://www.eecs.berkeley.edu/~pister/SmartDust.

Since a MEMS visibility is minimal, the military uses of smart dust has a lot of potential, especially in surveillance and the detection and monitoring of chemical or biological agents. Now envision smart dust motes equipped with legs and wings. Yes, those are in the works with the MFI (Mechanical Flying Insect) project being founded by DARPA. As with so many brilliant ideas, there is also a sinister side to smart dust. Imagine it mixed in with paint, in which entire walls are equipped with sensors, tracking your every move. Paranoia or possibility, time will tell. But Kris Pister points outs "As an engineer, or a scientist, or a hair stylist, everyone needs to evaluate what they do in terms of its positive and negative effect. If I thought that the negatives of working on this project were larger than or even comparable to the positives, I wouldn't be working on it. As it turns out, I think that the potential benefits of this technology far far outweigh the risks to personal privacy."[1]

What's the math behind this small stuff? Logic, ultra small numbers, programming, and problem solving all play important parts in MEMS and motes.

---

[1]http://www.eecs.berkeley.edu/~pister/

Starcage
and the works of Akio Hizume

What do the mathematical concepts of lattice theory, Penrose tiling, the golden mean, the Fibonacci sequence, pentagonal symmetries and quasi-crystal geometry have in common? They are major players in the works of Japanese architect Akio Hizume. His genius and imagination combine architecture and mathematics to create exciting new shapes which reflect his fascination with structures found in mathematics and nature. As he says, "I don't separate both science and art. Both are human arts."[1] As a result, his sculptures, architecture, and music evolve from these mathematical ideas.

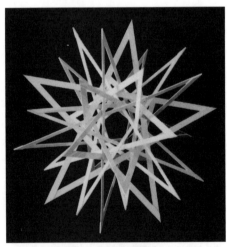

The Starcage, PLEIADES ©1995 Akio Hizume. It consists of 6 pentagrams and 30 plastic rods. It is commercially available at starcage@mbb.nifty.com)

Imagine a group of congruent pentagrams, each made from 5 rods, held together without the use of any adhesive, wires or strings, but by the tension created by their interwoven parts. The pentagrams lie flat against one another until they are hit or tossed onto a flat surface. Then, as if by magic, a 3-dimensional geometric shape emerges. In 1999, Hizume created a Starcage[2] consisting of 180 rods. His design won the

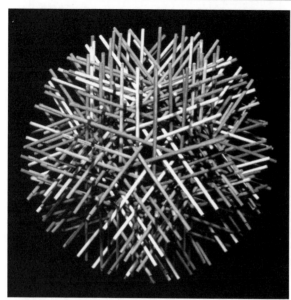

Starcage: MU-MAGARI No. 5 © 1999 Akio Hizume. It consists of 180 rods also designed around the symmetry of a dodecahedron using quasi-periodic patterns.

Silver Prize at the International Design Competition in Osaka, Japan. His MU-MAGARI[3] Starcage also consists of 180 rods and is designed around the symmetry of a dodecahedron using quasi-periodic patterns. All his Starcages are totally self-supporting. In fact, he has even created a self-standing Starcage (BAMBOO HENGE No. 5), which allows people to enter into its center. Hizume uses bamboo rods to make his Starcages.

Utilizing the golden mean and Fibonacci numbers, Hizume composed Fibonacci Kecak—music consisting of only 9 periodic rhythms, which repeat every 2000 years! (You can hear a 7 minute clip from his piece by logging onto: http://homepage1.nifty.com/starcage/fibonaccikck.html)

A *lattice* is an infinite array of points, spaced so that any point can be shifted onto any other point in the lattice by the arrangement of symmetry. The points defined by the x,y-coordinates of integers in a Cartesian plane is an example of a lattice. Among other types of lattices are those found in crystallography.

Tiling in mathematics is also called tessellating, which is covering a plane with a particular shape or shapes so that no gaps are left. The diagrams show how congruent equilateral triangles, squares or hexagons can be used to tile a plane such as a floor. The vertices match perfectly leaving no holes.

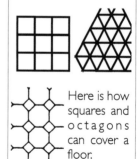

Here is how squares and octagons can cover a floor.

These examples are known as *regular periodic tilings*, here the design repeats on a regular basis as the eye moves vertically or horizontally. A pattern is not repeated in *nonperiodic tilings*. For example, consider tiling with stag- gering squares.

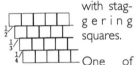

One of the most famous nonperiodic set of tiles is the Penrose tiles, composed of

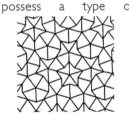

just two shapes, a dart and a kite. Penrose tiles possess a type of

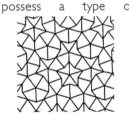

symmetry called *fivefold (rotational) symmetry* which means a tiling pattern can be matched up to another on the plane after it is rotated 1/5 of the way around, as can be done with a pentagram. Penrose tiles also have *tenfold symmetry*.

Two flat objects are symmetrical to one another if they can be made to match up when they are

In 1997 he was commissioned[4] to design the *Democracy Steps* for Cedar Falls, Ohio. He specifically designed the descending pathway of steps, which reflects mathematical principles of the Fibonacci sequence and a one-dimensional Penrose lattice, so they would lead to one of Ohio's most beautiful waterfalls. The individual steps are varied so that the walker alternates the leading foot and establishes a comfortable rhythm. Hizume designed *Democratic Steps* to be as effortless as possible, thereby making it feasible for almost any walker to experience art in a public space. In addition, *Democracy Steps* lets the walker focus on and enjoy the

Democracy Steps. ©1997 Akio Hizume

46

Hizume's BAMBOO HENGE No.5 . © 1998 Akio Hizume. Nakano, Tokyo, Japan. Photo: T. Ninomiya.

natural surroundings of the walk rather than having to concentrate on the effort or steps of the walk.

In his works of "neuro-architecture", Hizume draws on such mathematical ideas as Penrose tiling. His design illustrates an experimental city. Hizume feels *"there is an essential power in architecture to educate people and to create more freedom in and for them. Many museums are rectilinear, with square rooms, and exhibits are arranged chronologically. However, in neuro-architecture, linear paths do not exist; people can access its spaces randomly. They may, at first, become confused and perhaps even get lost within neuro-architecture, but* either reflected about a line, rotated about a point, or translated (moved) or glided in particular direction.

Quasi-crystals were discovered in 1982. Until this time, all crystals were considered periodic, i.e. composed of a periodic arrangement of identical polyhedron building blocks, and were considered a 3-D periodic tiling. In 1982 chemist Daniel Shechtmann found a way to produce a crystal that did not

Goetheanum 3 axonometric projections, exterior. ©1990 Akio Hizume. ink on paper 415x580 mm.

have 3,4, or 6-fold symmetry periodic tiling. In 1984, physicist Paul Steinhardt verified that these nonperiodic crystals possessed 5-fold symmetry, and he called them quasi-crystals.

The *golden mean* (also called the golden ratio and golden section) is the point on a line segment, A____B__C creating the following ratio (AC/AB) =(AB/BC). The golden mean appears in many shapes. Among the most popular shapes in which it appears are the golden rectangle (a golden rectangle can then be formed with side AC and AB) and the pentagram. The *Fibonacci numbers* 1,1,2,3,5,8,13,... first appeared as a solution to a

as they become more familiar with it, their minds will become educated and more advanced....In a sense, neuro-architecture is a two-dimensional arrangement of the one-dimensional Democracy Steps"[5.] As he points out, Penrose lattices appear in nature so why not in architectural designs. Utilizing one, two, and three-dimensional Penrose-lattices as a grid planning, he refers as to them as a *"self-similar and quasi-periodic city"* and the *Goetheanum 3* monument as a "six dimensional" structure because *"The architecture was designed based on six equivalent coordinate axes. It seems to be very complicated in a 3D world, but is very simple in a 6D world. The coordinate system is a shadow (a projection) of 6D on 3D space. We can only see the higher dimensional affair as shadow. But we can live there essentially."*[6]

Viewed from overhead one sees the quasi-periodic floor plan in its shapes, and from the side one senses its various dimensions. Although the feeling of such space may initially cause some disorientation, Hizume believes the overall effect will enhance the working of one's mind. The melding of his interests, fascination, and passion with forms found in nature, mathematics, music and art have a profound influence on his ever evolving architectural shapes.

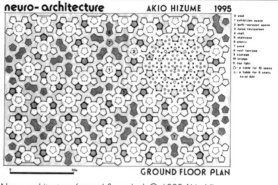

Neuro-architecture (ground floor plan). © 1995 Akio Hizume. ink and pencil on paper 200x300 mm

problem Fibonacci posed in his book *Liber Abaci* in 1202. Mathematicians have repeatedly found this sequence of numbers popping up in nature, art, and music. Each successive number of the sequence is generated by adding the two previous numbers. The golden mean is concealed in the Fibonacci numbers. The ratio of two consecutive Fibonacci numbers get closer and closer to the value of the golden mean, in fact its limit is the golden mean.

---

[1] Starcage wbesite: http://homepage/.nifty.com/starcage/index.html

[2] In1992 Hizume invented his 3-dimensional 6-axes self-supporting complex of rods named which he named starcage (Japan Patent Pending).

[3] Hizume describes MU-MAGARI as is self-complete, self-independent and self-supporting, which can be enlarged so that as it is made wider it becomes more symmetrical.

[4] The Hocking Hill State Park, Artists Organization o Columbus, Hocking County Tourism Association, and Ohio University-Lancaster's Wilkes Gallery brought Hizume ot Ohio.

[5] Ibid, footnote 1.

[6] From personal interview.

In search of AI
mathematics and artificial intelligence

Whether it was the R2D of *Star Wars*, Data or the Doctor from the *Star Trek series*, HAL from *2001: Space Odyssey*, or David and Gigolo Joe in AI, the idea of artificial intelligence has intrigued our imaginations, confused our existence, and in some cases sent chills down our spines. Today, the science of AI, seriously considers the reality and future possibilities and uses of robots designed for specific tasks. Many of us witnessed via television or the internet the exploration of Mars' surface by the Mars surveyor. Though not fashioned as a humanoid, the robot "landrover" was launched to gather data on Mars. We learned about the use of unmanned reconnaissance planes in the wars in Kosovo and Afghanistan. Naturally these "robots" are not classified as AI, yet they are steps leading to the evolution of AI. AI is moving in multiple directions simultaneously. We already have machines that recognize faces, distinguish voices, analyze conditions be they temperature, water levels, or braking systems. While none of these is classified as AI, some are dubbed smart machines.

AI, artificial intelligence: The ability to think and reason as a human would.

The phrase *artificial intelligence* was first coined in 1956 by John McCarthy of MIT at a summer conference[1] at Dartmouth, NH which focused on the research being done to make machines which behaved, performed, and processed information like humans. Robotics, automata, computer games, and smart machines all exhibit a form of intelligence. If a person were to instantaneously match a set of fingerprints, recognize a face, or identify a voice — these skills would be considered a form of intelligence. When a machine does this is it considered intelligent? In 1950 British mathematician Alan Turing proposed an AI test in which a computer could be called intelligent if it could fool a human into thinking it was human. The computer and the person would be in separate rooms, and would communicate via a computer keyboard, much in the way today's chat rooms are run. The person would type in questions to other people, but one of the recipients was really a computer. In 1990 this definition of AI was used by Dr. Hugh Loebner of the Cambridge Center for Behavioral Science. He offered a $100,000 grand prize for the computer whose responses could not be distinguished from those of a human. The machines only need to converse with the tester on a

An important step in developing AI is devising methods for processing and transmitting information. Today, the most widely used systems are *digital* or *analog*. *Analog signals* use electromagnetic waves, object, physical forces, or distances to measure and subscribe characteristics of an object and to transmit that information. There are many types and shapes of forms that analog systems use for their signals. On the other hand, *digital systems* use only numbers to process and transmit data. They use the binary number system with its strings of 0s and 1s to describe the characteristics and measurements of an object. These 0s and 1s can be easily transmitted via electricity with *0* assigned to *off* and *1* to *on*. If you have a watch that you wind, use a slide rule, take photos with a 35mm camera, listen to music on an LP— you're using analog devices. But if your watch has an LED display of numbers, you use a hand calculator, have a digital camera, or play an audio CD for music — you're into digital devices. In the past, telephones were only analog. Today there are telephones that are purely digital, analog, or a combination.

limited topic. A $2000 consolation prize and bronze metal is given annually to the creator of the "most human computer".

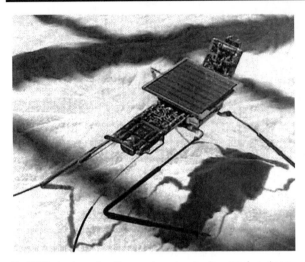

Mark Tilden's solar walker. Photograph courtesy Ian Bernstein./www.beam-online.com

One of the first mindless robots, *Homeostat,* was made by Ross Ashby in 1952. It consisted of a creature that could reorganize itself and stay alive even after some of its vacuum tubes were intentionally destroyed. But as with many intriguing things, Homeostat was before its time, and thought of as a curious but useless device. Today these mindless analog robots are no longer considered useless. They were first reintroduced by Rodney Brooks of MIT and now other scientists are exploring this area of AI.

These definitions of AI link intelligence to computers which are digital devices. Yet, analog machines are also being explored with fascinating success. Today, roboticist Mark Tilden at the Los Alamos National Laboratory in New Mexico has resurrected the idea of analog robots. Tilden was introduced to analog robotics in a talk given by Rodney Brooks of MIT in 1989. Ever since then he has been building and experimenting with analog robots. His robots take on a myriad of shapes and forms. Imagine a room full of insect-like robots. One that moves and feels like a snake. One that resembles a beetle. Imagine a robot with long wire legs, a flat head and blue eyes(sensitive to light, infrared, and heat) able to find you anywhere, even in the dark. Or *Roz* the 6' robot. Or the "self-aware" robot *I-machine* — the robot with a head within a head. Like his early analog robot, *Walkman,* they are able to traverse uncharted terrain and obstacles — retaining and reusing regained experience — all without a brain. They are amazing mechanical bug-like robots that move, overcome obstacles, and remember and apply what they experienced without computer processors. All these analog robots function without computer processors. How do they do it? Tilden explains that we ourselves are not rational animals, but "a solid core of pure chaos bounded by linear systems keeping us regulated toward some level of cohesiveness with our world. We are chaotic creatures who are made rational by our environment."[2] His work taps on this premise so that his robots present a "brand new way to think about thinking."[3] Their neurons are nonlinear elements whose sensors and mechanical structure are linear devices which maintain the robot in a state equilibrium. Tilden assembles his robots from transistors, capacitors, resistors, LEDs, sensors — stuff one would find in a transistor radio , a CD player, or tape deck. It is uncanny how a bunch of wires with a

> A *transistor* is an electrical device used to control the flow of electrons in a circuit.
>
> A *neuron* is an impulse conducting cell of the nervous system. The impulse is in the form of an electrical charge initiated by a stimulus.

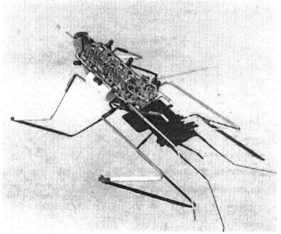

Mark Tilden's unibug. Photograph courtesy Ian Bernstein./www.beam-online.com

Mark Tilden's sidewinder. Photograph courtesy Ian Bernstein./www.beam-online.com

few transistor neurons powered by solar energy suddenly begin moving. Their goal is survival while always maintaining and seeking better power sources. Tilden puts his robots through their paces. Once a robot has painstakingly overcome an obstacle, Tilden makes it confront the obstacle again seeing if, when and how it "remembers" what it just learned. He may purposely disrupt wires to observe how it will adapt with a disrupted setup.

Tilden determined how to use analog circuits in a *complex nonlinear way*. There is no programming. The robots function on an innate survival instinct, which makes it possible for them to function in an unstructured unspecified world. The beauty of these analog robots is that they are inexpensive to make and experiment on, and are solar powered. Furthermore, once you've found out what you wanted from a design you do not have to hesitate about moving on to a new prototype and next generation. Most important, their wires, transistors, neuron net possess an almost uncanny system which exhibits *nonlinear, complex* and *chaotic behavior*[4].

Among Tilden's innovations we find his patented *two-transistor neuron*, which is a loop of oscillators with chaotic elements, and his *unicore structure* which allows the robot to retain patterns of stimuli and function as its memory. This *"neural net brain"* is not a digital computer but an additional analog layer which lets its nervous net body to retain rhythms. Tilden points out that "With the unicore, you can actually remember, and move, and then close that loop so you have a structure which is pretty close to being an artificial life-form."[5]

The **H**euristically programmed **A**lgorithmic **C**omputer a.k.a. HAL of *2001: Space Odyssey* was a fictional digital computer modeled after the 1980s pioneering work of neural networks of scientists Marvin Minsky of MIT and other researchers. They were looking for ways to automatically generate neural networks, thereby making them able to self-replicate, i.e. grow artificial brains. Arthur C. Clarke, the author of *2001: Space Odyssey*, predicted in a 2001 interview that "any sufficiently advanced technology is indistinguishable from magic; as technology advances it creates magic, and (AI is) going to be one of them....When nanotechnology is fully developed, they're going to churn (artificial brains) out as fast as they like."[6]

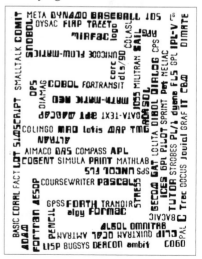

Digital AI depends on communicating and instructing the computer. Every figurative and actual step it takes has to be spelled out for it via computer language. Over the years scores of languages have been developed to handle various jobs, witness the diagram of a collage of computer languages. An analog AI can be influenced by how it is constructed, but it ends up performing according to how its layers of analog components react to environmental stimuli, rather than what an algorithm has outlined.

The pet robot Aibo developed by Sony.

Things that exhibit forms of AI are being constantly developed, tested and marketed. We find robot dogs, pets, and toys becoming a big hit, as witness the popularity of Sony's robot dog Aibo[7]. Smart machines are being widely used in the home and workplace[8]. Advanced devices for biometrics are becoming more widespread. The medical world employs smart diagnostic tools or future nanobots as operating tools. Answering systems are getting people accustomed to artificial voices. Today the military uses remote controlled devices to carry out missions, such as unmanned drone planes, submarines, and the Martian surveyor.

AI will play a major role in DOD(Department of Defense) and DARPA(Defense Advance Research Projects Agency) — military research projects which include robots to handle dangerous situations such as bomb disposal, swat missions, gas or chemical leaks, finding land mines, environmental toxic clean-ups, rescue missions, mini submarines, and reconnaissance missions.

Where does mathematics come into the picture? Just about everywhere. We see the roles that non-linear mathematics, nanotechnology, complexity theory, and chaos theory are playing in analog robotics. We know that without *Boolean algebra* and logic the first electronic computers would not have been possible. Without math one could not communicate with a computer. The binary systems held the key to converting electrical current to computer language[9] for digital computers. What about all the computer program languages that have evolved over the years in efforts to instruct these early machines to do tasks? The task of trying to get machines to reason, think, and apply what they know is what digital AI is about. Mathematicians are at work trying to integrate computer and *fuzzy logic*. *Informational geometry*, which studies how information is processed and communicated by the brain, is another important area for AI. We know that an enormous supercomputer whose abilities range from analyzing humongous quanti-

Boolean algebra was developed by George Boole in 1847. His algebraic system can reduce logical statements to a symbolic language in which arriving at a conclusion or reasoning seems as mechanical as solving an algebraic equation. Logical statements are translated into elements of binary operations. In addition to computer science, Boolean algebra is also used in probability and topology.

Fuzzy logic is not true or false logic (mathematically described by the numbers 1 and 0), but rather can consider the infinite realm of possibilities (everything between 1 and 0). For example, if you could ask a fuzzy logic robot "Is it raining?", rather than just taking the question literally and answering yes or no, it might reply "not yet" or "the chance of rain is 20% with partly cloudy skies" or some other answer predicated upon accessible information.

Today many researchers are pursuing different avenues in search of AI by designing smart software with practical applications. These programs rely on *Bayesian probability* models based on *Bayes's theorem* which was developed by T. Bayes in the mid-1700s. The theorem deals with *conditional probability* — the probability of an event taking place based on the conditions of other event(s) occurring.

These Bayesian models are able to scan and evaluate factors and data. They can prioritize and sort information, such as telephone messages, e-mail, and appointments, which are designed specifically around the user's needs and habits. In much the same way certain programs analyze people's buying habits in order to target consumers with spam ads, junk mail, or messages. This new focus for AI does not seek to replicate human intelligence, but to design systems which help enhance human abilities. As computer scientist Patrick Winston of MIT says "We don't try to replace human intelligence, but complement it."[10]

ties of information in seconds (e.g. the IBM supercomputer, Deep Blue, which was programmed to analyze 200 million chess positions a second and defeated world chess champion Gary Kasparov by using brute computing force rather than ingenuity) does not quality as AI. The quest to program digital computers to think as humans has led to many things: (1) the exploration of using *neural networks*, which try to introduce the simulations of the same physical connections found in human brains); (2) experimenting with *expert systems*, which are computer applications that make decisions in real-life situations); (3) programming computers to interact in the user's language has been limited to a finite set of recognitions rather then understanding and interpreting the meaning of the words; (4) programming devices to react to what they see, hear and sense; (5) working with *genetic algorithms*; (6) using nanotechnology.

The future of AI may lie with a combination of analog and digital devices. There may not be enough memory to instruct a digital computer to be able anticipate all possible interactions and situations with humans and its environment and function as a human brain, let alone attaching body parts to simulate the intricate, complicated and infinitely varied movements of a human. Perhaps if digital and analog forces are joined with ever advancing areas of mathematics and science — AI will evolve.

---

[1]John McCarthy, Marvin Minsky, Nathaniel Rochester, and Claude Shannon are often referred as the founders of AI.

2Trachman, Paul. Redefining Robots, SMITHSONIAN, February, 2000.

[3] Hapgood, Fred. Chaotic Robots, WIRED, September, 1994.

[4]See chapter on chaos for definitions.

[5] Ibid. Footnote 3.

[6]Cyber View Scientific American Jan 2001.

[7]Aibo is a robotic pet developed by Sony. The owner, using specific software decides — if its a dog or cat— whether its male or female— its age. Originally selling for $2,500, the new generation sells for $1,500 with add on software that allows it to take pictures and the ability to respond to fifty voice commands.

[8]Among these machines we find smart washing machines with sensors telling water levels and agitation needed for the soiled garments or to repeat a wash cycle, smart thermostats with fuzzy controllers, smart vacuum cleaners which adjust the suction according to its sensor readings, and smart software using *Bayesian probability* models to prioritize and sort telephone messages, e-mail, and keep track of appointments.

[9]LISP or Prolog are examples of programming languages specifically developed for AI. They are referred to as symbolic processing languages, as opposed to numerical processing languages such as Fortran, COBOL, Basic, or C. The symbolic processing languages codify information into facts and rules or procedures for handling it.

[10] *A.I. Reboots* by Michael Hiltzik, page 50 *Technology Review*, March, 2002

art in nature, nature in art

Andy Goldsworthy's art

A ndy Goldsworthy's palettes are objects found in nature—sticks, stones, ice, water, leaves, sand, snow, moss, and pebbles. He draws inspiration as he wanders in a forest, along a coastline, through a countryside, over a meadow, and even at the North Pole. The entire Earth is his canvas and nature his model — her shapes and forms inspire him. Goldsworthy's works are a marriage of art and nature. Time and the elements play important parts in his art, and although most of his art is temporary, it mirrors all things in life—it is constantly evolving and changing. As he says, "Movement, change, light, growth and decay are the lifeblood of nature, the energies that I try to tap through my work. ...Nature is in a state of change and that change is the key to understanding. I want my art to be sensitive and alert to changes in material, season

Arche. Courtesy of Sculpture at Goodwood
website: www.sculpture.org.uk.

and weather. Each work grows, stays, decays. Process and decay are implicit. Transience in my work reflects what I find in nature."[1]

Immediately after completing a work, he documents it with photographs, and from then on nature takes over. His works resemble familiar shapes and forms which appear and reappear in nature— shapes and forms which mathematicians have used and described with equations and graphs and in theories.

In his work *Broken Pebbles*, we see the *equiangular spiral* — a natural growth form that lends itself to becoming continually larger and continually smaller simultaneously. We know the shape. We've seen it in ammonites, the nautilus, fingerprints, and the web of an orb spider. It is a beautifully familiar shape appearing in the golden rectangle and tied to the golden mean. In his *Soul of a Tree*, we catch a glimpse of a 3-dimensional icicle spiral, a helix wrapped around a tree's trunk. Goldsworthy points out that with each

The *equiangular spiral* is also known as the logarithmic spiral, which was discovered by René Descartes in 1638. Over the years, mathematicians have discovered its interesting characteristics. For example, any of its radii which pass through its origin are cut at the same angle. Like a fractal, the equiangular spiral is self-similar, meaning any portion is identical to another portion of the same curve. Another fascinating characteristic is that if the spiral is rolled out along a level surface, the path traced by its center as it unwraps is also a straight line.

work "I want to get under the surface. When I work with a leaf, rock, stick, it is not just that material in itself, it is an opening into the processes of life within and around it. When I leave it, these processes continue."

Stone River (2001), by British artist Andy Goldsworthy, shows the meandering of a river.
A gift from the Robert & Ruth Halperin Foundation to Stanford University.

His work of a pool of beech leaves (see *Stone* by Andy Goldsworthy, Harry N. Abrams Publishers, New York, 1994.) captures the momentary lull of trapped water while his *Stone River*(2001) shows the meandering of a river.

Other mathematical ideas occurring in his works include—arches, fractals in an arrangement of goose feathers, pyramids stacked in the gathered stones series, triple junction in the cracks of *Hard earth 1992*, and tessellations in the slate floor of *Stone sky 1992*.

Almost all rivers and streams *meander* in the same manner. Why does a river flow the way it does? Is it responding to the shape of the land its transversing? Not unless there are extreme slopes. Mathematicians and scientists have observed that water does not flow in a straight line downhill. It meanders back and forth apparently trying to avoid a direct path down. The winding, in fact, is predictable and not tied into the land's topography. Its curves are elliptic integrals, curves which are extremely smooth which exhibit the least change in their curvature's direction. What patterns have been observed? • no river runs in a straight path for more than ten times its width • more often than not the radius of a bend is usually 2 to 3 times the width • the wavelength of analogous points of analogous bends is 7 to 10 times the width. There are various scientific theories for why rivers flow in this way. Perhaps, it is due to centrifugal force; perhaps, it is caused by uniform expenditure of energy; or perhaps it is the self-correcting characteristic of a complex system.

---

[1] http://cgee.hamline.edu/see/goldsworthy/see_an_andy.html

Do bees count?
bees and spatial coordinates

D o bees count? Nature has many mathematical surprises and bees are one of them. Bees instinctively use the hexagon, which gives the most area for the least material, to design the cells for their honeycomb. Bees also communicate through "dance" movements, which can be thought of as codes. The melipona bees (the stingless bee found in the tropics) is able to communicate distances in 3-dimensions. Since their food sources are located both on the floor and canopy of the jungles, these bees must be able to describe spatial directions, whereas the communication of honeybees only deal with planar coordinates, since they forage predominately near the ground. How does the melipona give its directions? By buzzes and turns. The height is indicated by the length of the buzzes. Bees decipher these sounds by feeling their vibrations, since they do not have eardrums. Distance is communicated by using a series of semi-

circular turns and buzzes. It is unclear whether the turns may also indicate when a new subject is being introduced. It seems that the longer buzzes indicate longer distance.

e-paper & mathematics

ho would have imagined a few centuries ago that the numbers 0 and 1 would play such prominent roles in our everyday lives. From these two numbers, mathematicians devised the binary number system, which became the means by which electricity could be harnessed to communicate and direct computer functions. Yes, strings of 0s and 1s in billions of combinations produce the words you type, the graphics artists create, the problems scientists solve, the analyses

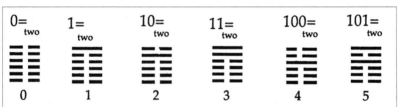

Mathematician Gottfried Wilhelm Leibniz and Pére Joachim Bouvet, a Jesuit missionary in China, corresponded from 1697 to 1702. It was through this correspondence that Leibniz learned that the above I Ching hexagrams were connected to the binary numeration system he devised. He noticed that if he replaced a broken segment with a 0 and a solid segment with a 1, the hexagrams were binary numbers. Both Leibniz and Bouvet felt that the Chinese had discovered the binary system through I Ching (circa 8 B.C.), but there is no evidence to indicate this.

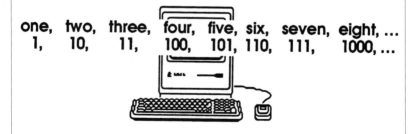

e-paper —*The next genera-tion of the printed word — from clay tablets to papyrus to paper to the Gutenberg press to computer word pro-cessing to electronic books to e-paper.*

computers are directed to carry out. The power unleashed by these two numbers along with Boolean algebra is mind boggling. But, in another decade we may witness yet another amazing feat 0s and 1s will tackle — *a transformation of the printed word by allowing it to reach us by wireless modes.* Our daily newspaper may not have to be delivered by the papercarrier or read off a computer screen, but instead be transmitted to special paper, paper that looks and feels like newsprint, but paper that is coated with a special ink that changes every morning to your daily newspaper via wireless communication.

Boolean algebra is a mathemati-cal system developed by George Boole in 1854, which applies an algebraic framework to the thought process. His work entitled *An investigation of the laws of thought, on which are founded the mathematical theories of logic and probabilities* was the first form of symbolic logic developed, though different from that developed later by Gottlob Frege and Giuseppe Peano.

In Boole's system, 0s and 1s are the only values variables can assume, and functions are expressed by using *and, or,* and *not.* Boolean algebras have been used in switching theory, logic design and other computer science applications.

At the basis of e-paper is the develop-ment of electronic ink. The idea of elec-tronic ink first occurred to Nick Sheridon in the early 1970s when looking for new alternatives for computer monitors. While at the Palo Alto Research Center (PARC) at Xerox corporation he developed microscopic balls that were half black and half white. When an electronic charge was applied to these microscopic balls, light and dark images were created as they rotated from the charge. Within a few years Sheridon was able get these microscopic balls to form the letter "X"(standing for Xerox). Excitement over such innovative technology did not surface for another twenty years, however. Eventually the concept of e-paper captured the imagination of other scientists. In 1997 physicist Joseph Jacobson, mathematician Barrett Comiskey and mechanical engineer J.D. Albert, Russ Wilcox, and Jerome Rubin founded E Ink. Since then various companies have been collaborating with E Ink. These include IBM, Motorola, Hearst (magazine and newspaper publisher), and Lucent Technologies. 3M and Xerox are working independently on their own designs for e-paper.

E Ink's technology also uses millions of microscopic balls which are filled with light and dark dyes that change color when an electronic charge is applied. In addition, Lucent's laboratories have developed flexible transistors which are capable of being printed on very thin sheets of plastic paper. The production process of e-paper will resemble that of printing rather than the silicon process used for LCDs (liquid crystal displays). LCD displays require a constant flow of electrical power. The beauty of e-paper is that once an image appears on it no additional power is needed to sustain it until the image is changed, and thus it uses very little

When e-newspaper becomes a reality, it will feel just like our present day newspaper. Its beauty will be that it never will need to be recycled.

power. The ink, consisting of millions of minute microcapsules is liquid. Each capsule contains a white particle that is floating in dark dye. When a charge is applied, the capsule's white particle moves to one side and becomes visible making the surface appear white at that particular location. When an opposite electric field is applied, the white particle is forced to the other side of the capsule where it is hidden in the dye, thereby making that particular area of the surface appear dark.

What are the advantages of e-paper? When one views an LCD, the quality of the image usually depends on the angle from which you are viewing the screen. For e-paper, on the other hand, a good image is given from all locations as if it were plain printed paper. In addition, e-paper can be read in direct sunlight and most other lighting conditions, unlike LCD which requires backlight. E-paper is much lighter than LCDs, can be printed on plastic, glass and metal, and is easily adapted to any size format without a significant increase in cost.

Where is e-paper technology now? E-paper sandwich board signs have gone up in various cities across the USA promoting shopping, services and specials. On a smaller size scale, the technology will probably also be used on prices tags for items in stores. When the prices change, new information will be downloaded onto the tags. And eventually e-paper will be used for our daily newspaper. We may also see e-paper technology adapted and used in clothing, food and drink containers, cups, shoes— anything that can carry the printed word.

The April 24th *Proceedings of the National Academy of Sciences* (2001) reported that Lucent Technologies and E Ink devised a way to produce e-paper using simple inexpensive printing methods. "The real dream has been to have electronic newspapers or electronic books that are manufactured in a way that you would manufacture a regular book...This is the first time that anybody has manufactured all the elements — both the electronics and the display itself — by printing,"[1] said Joseph Jacobson.

Senior vice president of Hearst Interactive Media, Kenneth Brofin, pointed out that E Ink's technology "Stretches across everything we do. If you look out five or ten years, this technol-

The earliest paper-like form for writing was the Egyptian papyrus scroll. The papyrus plant was stripped and pressed to create "paper". The word paper originated from the word papyrus, even though paper, as we know it, was not invented until 105 A.D. in China. This section of a scroll is a rendition from the Rhind mathematical papyrus, circa 16th century B.C..

ogy could be very, very real and change the way people get information."[2]

There are predictions that our bookshelves will take on a new look. Rather than rows and rows of books, only a handful of books will suffice. The publisher will download their current best seller on an unused blank e-paper book, and the reader will enjoy it just as if it were a real book. You will not have to fetch your newspaper every morning; rather, a permanent e-newspaper will be updated everyday via a wireless broadcast. Too far fetched? The technology is there evolving. At this time, one is left speculating about both the physical and environmental impact of such technology.

---

[1]*New device opens next chapter on E-paper.* by J. Gorman. Science News, vol.159. April 28, 2001.

[2]*E Ink writes its future on e-paper.* by Alec Klein WSJ Interactive Edition, January 4, 2000.

mathematics mints a coin

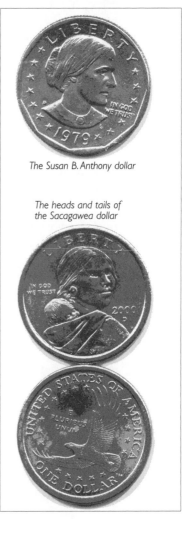

The Susan B. Anthony dollar

The heads and tails of the Sacagawea dollar

When it came to minting the new Sacagawea dollar coin, it was not as simple as stamping a new design on a coin. The new dollar had to match the specifications of the Susan B. Anthony dollar in order to work in existing vending machines. The dimensions — size, shape and weight — were easy to satisfy. The trick was designing an alloy which had a gold luster and the same electromagnetic signature as the Susan B. Anthony dollar. Why? Because vending machines, in addition to checking the size and weight of a coin, also check the electric conductivity of the metal. Not being able to get a match would mean revamping the existing vending machines — a formidable expense. Fortunately, after over

30,000 test samples, Olin Brass company of Illinois (which has been supplying the mint with alloys since 1964) developed an alloy which had the same size, shape, weight and conductivity as the Susan B. Anthony dollar. The new alloy ended up being 77% copper, 12% zinc, 7% manganese and 4% nickel.

Could some *linear algebra* have been used because of the various conditions and restrictions which had to be met?

Linear algebra is a field of mathematics which uses linear expressions, such as $x+2y-5z-0.8m>15$ to solve problems which have various conditions and restrictions. An expression is written for each condition that must be satisfied from the given information. Then a set of possible solutions is generated, often using graphing techniques. The next step is to find the optimum solution, which in some cases may be the one that uses the least amount of materials or costs less.

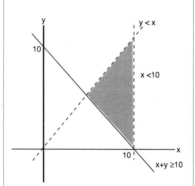

The shaded region indicates possible solutions for a linear algebra problem with three expressions.

the sound of
mathematics
music & mathematics

I n ancient times music was treated more like a science than an art. Over the centuries mathematics has been used to describe, dissect and analyze music. Fractions and ratios play an important role in music and score writing. The lengths of musical notes are identified according to fractions — fractions of the geometric series (1, 1/2, 1/4, 1/8. 1/32. 1/64). There is the whole note, the half note, the quarter note , the eighth note, the 1/16 note, the 1/32 note, and the 1/64 note. Fractions also designate music's time signature, the number which appears at the beginning of a musical score. Here

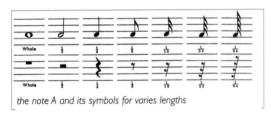

the note A and its symbols for varies lengths

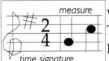

we may find the fractions 4/4, 3/4, 5/4, 6/8, or 2/2. The denominators of these fractions specify what length note (a quarter, eighth, or a half) is to receive one beat. The numerators, on the other hand, indicate how many beats there are in a measure. If one were to take a bar of notes and keep the same notes but change their lengths, a new "tune" emerges. The count value of each note changes the rhythm of a melody, and thereby changes the melody. Counting, tempo, and note lengths are but a few mathematical characteristics found in the musical score. Fractions or ratios have also been used to describe other characteristics of musical sound. Around 540 B.C. the Greek mathematician Pythagoras, working with a *monochord,* discovered the relationship between the vibrations of a plucked string and its length. He found a plucked string produced a certain note. For example, if a string produced the note C, then that string depressed at half its length produced the same sound,C, but an octave higher. In this way the Pythagoreans discovered that all the notes of a scale were related to integral ratios. For example, starting with a string that produced the note C, then 16/15 of C's length produced B and 6/5 of C's produced A, 4/3 C's gives G, 3/2 of C's is F, 8/5 of C's is E, 16/9 of C's gives D, and 2/1 of C's produces low C. In addition, the string's vibrations inversely conformed to these ratios. C_L(low C) makes 264 vibrations per second, but when the string is depressed at half its length the C_H is produced one octave higher than C_L and vibrates at 528 vibrations per second, which illustrates their lengths are inversely proportional to the frequency of their vibrations.

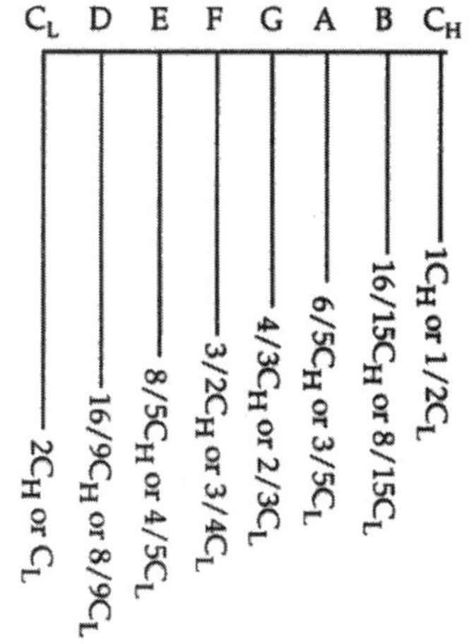

C_L D E F G A B C_H
$2C_H$ or C_L
$16/9C_H$ or $8/9C_L$
$8/5C_H$ or $4/5C_L$
$3/2C_H$ or $3/4C_L$
$4/3C_H$ or $2/3C_L$
$6/5C_H$ or $3/5C_L$
$16/15C_H$ or $8/15C_L$
$1C_H$ or $1/2C_L$

This diagram shows that as the length of the string, C_L or C_H, is modified by the integral ratios shown the notes of the diatonic scale are produced.

The ancient Greeks used letters of their alphabet to represent the seven notes of their scale. The Chinese used the five note pentatonic scale. In India music is mainly improvised with specific boundaries defined by ragas. The octave is divided into 66 intervals called srutis, although in practice there are only 22 srutis from which two basic seven-note scales are formed. The Persians divided their scale into either 17 or 22 notes.

A *monochord* is a single stringed instrument.

With the Pythagorean ratios the 7 octaves and the 12 fifths do not match perfectly, and the slight discrepancy is called the *Pythagorean comma*. When seven octaves are produced using the Pythagorean note ratios, the interval after the seventh octave does not return exactly to the first tone. To envision this imagine winding the seven octaves around in the shape of a helix, then the eighth C ends up a little passed the first C.

Consider the diagram of a piano key board showing 7 octaves. G_1 is called the first fifth of the scale with its Pythagorean ratio is 2/3. An octave is represented by the ratio 2/1 or 2. As the keyboard illustrates, 12 fifths should fit perfectly into the 7 octaves. But, for the Pythagorean division of notes on the diatonic scale we have:

7 octaves=2^7=128 which should equal 12 fifths, but

12 fifths=$(3/2)^{12}$ =531441/4096 =129.7463379. The two are close, but not the same.

The ratio
$(3/2)^{12} / 2^7 = 1.01364326\ldots$ is the value of the *Pythagorean comma*,

Namely, the lengths $C_L/C_H = 2/1$ while *[the frequency of (C_L)][(the frequency of C_H)* =264/528=1/2. The pitch of a note is determined by the frequency with which the string vibrates. The tuning of an instrument involves creating intervals between notes. In the case of the *Pythagorean diatonic scale* the intervals are integral ratios. In fact, the tuning of any scale on a stringed instrument relies on ratios and the length and tension of strings. The Pythagorean note ratios produce the diatonic scale, which has been used for nearly 2000 years in the western world, and its tuning has come to be referred to as the *Pythagorean tuning*. For over 2000 years it has been the tuning of choice. Pythagoras knew his system of dividing the octave using these geometrically perfect ratios also produced a slight discrepancy, but because the Pythagoreans felt only whole numbers and whole number ratios were the essence of everything, whole number ratios were preferred to introducing non-rational numbers into the ratios.

This glitch has come to be referred to as the *Pythagorean comma*. The problem with the Pythagorean comma can be heard when notes are played simultaneously — the slight discrepancy occurs in the tuning of certain notes in high and low octaves. When these are played simultaneously they make a sound that seems out of tune. The Pythagorean comma was first studied in 11th century China and later in 17th century Europe. Exactly equal intervals in a 12 note scale were explored, which today is referred as an *equal tempered* scale. As early as the 12th century musicians began to experiment with new forms and styles of music which departed from the Pythagorean ratios and tuning. In 1584 Chu Tsai-Yü introduced *equal temperament* to the Chinese scale, but this system was rejected in 1712 by the Ch'ing Dynasty for both traditional and religious reasons. In the 1600s both Galileo Galilei and Marin Mersenne worked on equal temperament, and in 1636 Mersenne described equal temperament in his work *Harmonie Universelle* (Universal Harmony) . He deter-

C	C#	D	D#	E	F	F#	G	G#	A	A#	B	C
1	$2^{\frac{1}{12}}$	$2^{\frac{2}{12}}$	$2^{\frac{3}{12}}$	$2^{\frac{4}{12}}$	$2^{\frac{5}{12}}$	$2^{\frac{6}{12}}$	$2^{\frac{7}{12}}$	$2^{\frac{8}{12}}$	$2^{\frac{9}{12}}$	$2^{\frac{10}{12}}$	$2^{\frac{11}{12}}$	$2^{\frac{12}{12}} = 2$

indicating the amount by how much the 12 fifths passes the 7 octaves. Doing a similar process on a monochord, beginning with low C_1 and moving up to C_8, the same discrepancy occurs.

Today musicians and composers predominantly use the equal tempered scale. Its octave is also divided into 12 equal parts and the frequency ratio between half notes, e.g. A to A# or B to B flat is determined to be 1.059...,
which is

$$\sqrt[12]{2} = 2^{\frac{1}{12}}$$

All 12 notes' frequency are distributed according to this irrational number.

Now consider once more the diagram of a piano key board with 7 octaves. Its G_1 or first fifth of the scale is identified by $2^{(7/12)}$. The octave again is represented by the ratio 2/1 or 2. Now calculating 12 fifths over the 7 octaves we get 7 octaves $= 2^7 = 128$ and 12 fifths $= (2^{(7/12)})^{12} = 2^7 = 128$. There is no discrepancy after the 7th octave.

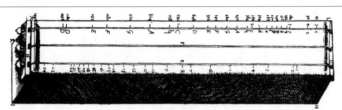

A rendition of the monochord illustrated in Mersenne's book *Harmonie Universelle*.
The top row of numbers indicate vibrations per second to the notes in the row below.

mined a way to relate pitch to a specific number of vibrations per second. Johann Sebastian Bach (1685-1750) was one of the first major composers and musicians to use the equal temperament in such works as *The Well-Tempered Clavichord*.

In the 19th century mathematician John Fourier made a new connection between mathematics and sound when he showed that all sounds —instrumental , vocal or any form of vibrations — were related to waves and their nature regardless of whether the waves were produced in water, earth, or air. He found that sounds could be described by mathematical expressions which happened to be the sum of simple *periodic sinus functions*. [These sinusoidal graphs illustrate that every sound's *pitch* is related to the frequency of its curve, its *loudness* to its amplitude, and its *quality* to the shape of the periodic function. In the last decade, mathematicians carried Fourier's work further by studying miniature sinus curves, called *wavelets*. It was the work of mathematician Ingrid Daubchies with these miniaturized sinus curves that linked the wavelets to time intervals varying according to the pitch of the sound. Today wavelet theory, joined with computer technology, can isolate and

Periodic function is a function which repeats its shape at equal intervals. A *periodic sinus function* is also called a *sinusoidal function* and uses the sine curve when graphed. Like the sine curve it is periodic, as shown in its shape below.

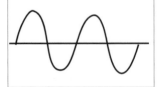

describe a sound in a conglomeration of noise, thereby making mathematics of sound essential to evolving voice recognition software and sound reproduction.

Acoustics is another area where mathematics played an important role in sound generation. Around the turn of the 19th century, Wallace Clement Sabine, a mathematics professor at Harvard, was asked to see if he could find a way to improve the abysmal acoustics in the newly opened Fogg Art Museum. He identified the problem as excessive reverberation and determined a way to quantify the reverberations by measuring the time it took, after a sound has stopped, for the reverberation of a sound to be barely audible. Today acoustic architects also take into consideration the type of music or noise and the size of the room when quantifying reverberation time. Sabines mathematical work helped him devise a simple inexpensive solution to the Fogg Hall problem. He had felt pads placed on certain walls to reduce the reverberations. The reverberation time helped him determine the optimum placement for the padding. Over the last forty years scientific studies using various input devices and computer modeling have helped in determining modifications that could be made in architectural design to improve acoustics of buildings and auditoriums. Mathematical analysis and measurements have shown that it was important to keep reverberation times below 2.2 seconds, that narrow halls worked better than wide ones, and the reflection devices such as ceiling hangings help regulate reverberation time and can be used to mix and diffuse the sound. Even so, the quality and reproduction of sound and music can be significantly

> The acoustics in the Louise M. Davies Symphony Hall in San Francisco were initially so poor that reflectors hanging from its ceiling were necessary to improve the acoustics .

> Manfed Schroeder, using number theory and quadratic residues, demonstrated how narrow wells designed into a ceiling could improve acoustics by diffusing and distributing sound throughout a hall.

improved with the introduction of computers and digital and analog replication devices. Computers, new software and improved peripherals have also been used to measure, analyze and improve the quality of musical instruments, and create synthesized music — all of which would not have been possible without mathematics.

While sound and music have been undergoing mathematization, musicians and composers have also used mathematics to create and formulate music by directly experimenting with mathematical patterns, algorithms and other mathematical concepts. This is not a new phenomenon. In the 1700s Wolfgang Amadeus Mozart used the rotational properties inherent in a musical score — a line of music rotated 180° produces a new and different line of music. For example, line (A) is the first line from a piece Mozart had written. Notice when it's rotated 180° it becomes line (B). So in Mozart's 12 line piece of music, (A) is the first last line played by a musician, but its rotated version (B) is the last line played by the second musician. Mozart designed the piece so

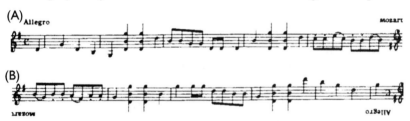

that musicians could play the score simultaneously — two sheets of music, each the rotated version of the other — beginning from opposite ends of the score. Since Mozart wrote an even number of lines in his piece[1], the two musicians never played the same line simultaneously. In this piece, the music Mozart wrote for both musicians fits perfectly together and works well as a duet.

In the 11th century, monk and early musical theorist Guido D'Arezzo was probably the first to actually write music using parallels lines to form a staff and name the notes of the scale. In addition he derived an algorithm(a sequential set of rules) to create a musical piece directly connected to a particular text. He first sequentially extracted the vowels as they appeared in the Latin text he planned to set to music. Then he devised a way to connect the text to music by arranging the vowels in the order they appeared in the text. He then matched the sequence of vowels to notes in two octaves, so that each vowel had at least three possible notes which could be selected, thereby allowing the composer some

> Here's how D'Arezzo's method worked. The two octaves & vowels could be listed:
> C D E F G A B C₁ D₁ E₁ C₁ G₁ A₁ B₁ C₁
> a e i o u a e i o u a e i o u
> This design allowed three possible notes to choose from for each vowel.

leeway in varied compositions. This example of the influence of mathematics on music composition is not an isolated one. In the 1600s Blaise Pascal and Pierre de Fermat launched the theory of probability using the game of dice to determine probabilities of the dice roll. Later some musicians were intrigued by this new area of mathematics and even introduced probability into some of their compositions. Mozart, in fact, wrote *Melody Dicer* based on a game of dice in which each measure of the piece depended on the outcome of a role of the dice. His algorithm associated the outcome of a role of dice — 2 through 12 — to a set of notes. Today many modern musicians and composers rely on computer software to generate hard copies of their musical scores and to write the various instrumental parts. There are even computer programs available for those who want to compose music in the style of a particular composer. But this is not unique to the 20th and 21st centuries. In 1775 Peter Welcker published *A Tabular System Whereby Any Person without the least knowledge of Music May Compose Ten Thousand Different Minuets in the Most Pleasing and Correct Manner* — the title says it all. Just as many artists exploit the computer as one of their mediums, a

The *golden mean* is a number generated from the ratio created by point B being placed at a specific location of segment AC

A   B   C

producing the ratio $|AC|/|AB| = |AB|/|BC|$. This ratio's value is $1.615...$ $=(1+\sqrt{5})/2 = \phi$. ϕ is the symbol used to denote the golden mean.

The *Fibonacci sequence* is the sequence
$1,1,2,3,5,8,13,21,...$
It's is interessting to note that ratio of consecutive Fibonacci numbers comes closer and closer to ϕ value.
$1/1, 2/1, 3/2, 5/3, 8/5 \longrightarrow 1.615....$

similar phenomena is taking place in music. "In a larger sense, nearly all if the music you hear today, both recorded and live, is electronic....the computer is inextricably woven into all stages of modern recording process: Even acoustic music such as string quartets and bluegrass is spliced and diced with all-purpose mixing software like *Pro Tools* and *Logic*. The wandering tones of mediocre(but remarkable) singers are routinely treated with pitch-correcting programs like *Antares Auto-Tune*."[2]

Compositions have also been based on such mathematical ideas as the Fibonacci numbers, the golden mean and fractals. Hungarian composer Béla Bartók (1881-1945) intentionally used Fibonacci numbers and the golden mean in some of his works. "Bartók based the entire structure of his music on the *golden mean* and *Fibonacci sequence* — from the largest elements of the whole piece whether symphony or sonata, to the movement, principal, and secondary themes and down to the smallest phrase."[3] In some of his works Bartók developed and used a Fibonacci scale. We find that physicist and modern composer Gyorgy Ligeti has deliberately used fractals in some of his compositions, as in his *Etudes for Piano*. Perhaps in

An *iterative algorithm* is a set of rules which inputs an initial value into a formula or applies a set of instruction to the value. The initial value produces a new value which is then also inputted into the algorithm to produce another value. The process is continued until all the required data has been generated. With an iterative algorithm there is no need to store the sound, only the initial value(s) and the algorithm.

the not too distant future some composers will use mathematical equations to store their music by reducing the composition to a one or more fractals described by *iterative algorithms*, such as those that have been used to compress digital art for storage and transmission.

Sometimes mathematics is described as the science of patterns. Music is also the repetition of patterns be they sequences of notes in a song or themes in a movement. Just as mathematics has evolved its own language and written notation, so has music. In music's case mathematics has influenced its written language, its structure and its creations, and continues to have profound effects on its evolving forms.

---

[1] For Mozart's entire piece see *Math & Music* by T.H. Garland and C.V. Kahn page 80.

[2] *Songs in the Key of F12* by Erik Davis. WIRED magazine, May 2002. Also see Siz Machines that changed the music world by Pat Blashill WIRED magazine, May 2002 for recent technological machines and their impact on the realm of popular music.

[3] From page 97 of *Connections* by Jay Kappraff, Mcgraw Hill, Inc., NY, 1991.

mathematics engineers your finances

Modern mathematics has branched out into many areas, and its tools are sought from all areas of society. In the banking world, many people no longer just save their money, they have it managed, or in some cases mismanaged. Financiers are looking to mathematics to help them achieve long term gains and insure against large losses. A mathematical model for pricing options was first developed in 1973 by Fisher Black and Myron Scholes. Using the mathematics of Brownian motion (which describes the movement of gas molecules in a closed chamber) the model likens the up and down movements of options to molecules of gas bumping and nudging one another. This model has proved invaluable in the area of mathematics and high finance.

In 1942 Itō Kiyoshi developed and

Itõ's Lemma

When a variable x follows
the Weiner Process, dx=ad+bdz,
G is a function of x & t follows

$$dG = \left(\frac{\partial G}{\partial x}a + \frac{\partial G}{\partial t} + \frac{1}{2}\frac{\partial^2 G}{\partial x^2} \right)dt + \frac{\partial G}{\partial x}bdx$$

proved a theorem now called Itõ's Lemma. The theorem gives a complicated mathematical equation which describes a random behavior, and therefore can deal with chaotic systems. In the 1970s, the United States abandoned the gold standard and international currencies entered a floating rate system. As a result, prices of currencies, interest rates, stocks, bonds, futures, options could fluctuate dramatically from moment to moment. At the same time, in the world of mathematics, interest was emerging in the study of random behavior, which ended up applying to the analysis and prediction of fluctuating prices of financial products. Itõ's mathematical equation directly applied to this evolving economics. In 1997, the Nobel Prize for Economics[1] was given to Robert C. Merton, in recognition of his contributions to the area of financial engineering. He had adapted Itõ's Lemma to produce a formula for price fluctuation.

In 1994 the Nobel Prize for economics was given to mathematician John Nash for his work on game theory[2]. In the 1950s Nash had come up with several concepts

Nash Equilibrium exists in certain types of games. For example, in a game in which no player can gain anything by changing his/her strategy if the other players keep their strategies unchanged. This game is said to possess a *Nash equilibrium* in its set of strategies and the corresponding payoffs.

Some games may have multiple Nash equilibria, while the Nash equilibrium does not apply in others.

which applied to game theory. One of which was the *Nash Equilibrium* which is widely used in game theory, and was the main criteria for his award, which read *"for their pioneering analysis of equilibria in the theory of non-cooperative games"*.

In the world of economics the use of mathematics is ever increasing — be it to make forecasts, to form statistical analyses, or even conduct what is now called financial engineering — a field which includes finance research and risk management. Today most financial institutions use some of the tools of financial engineering, and businesses seeking to manage your money utilizing financial engineering are available everywhere. A word of caution: Be certain the company you select knows its mathematics.

---

[1] Robert Merton shared his Nobel Prize with Myron S. Scholes *for new methods to determine the value of derivatives.*

[2] John Nash shared his Nobel Prize with John Harsanyi, Reinhard Selten *for their pioneering analysis of equilibria in the theory of non-cooperative games.*

numbers, numbers everywhere

Today there is no way to get away from numbers. Whoever first devised the concept of number could not have begun to imagine the impact of that seemingly simple concept. Over the millennia, mathematicians have gone far beyond those early counting numbers. In fact, their fertile imaginations led to ever evolving types of numbers. Many of us feel we are quite numerically sophisticated if we can expound on integers, rationals, irrationals, reals, complex, transfinite, transcendental, and imaginary numbers. But, in each of these sets more specific forms continually crop up. What about primes, composites, triangular, perfect numbers,…? There are many types of numbers named after the famous mathematicians who were responsible for them, such as Fibonacci numbers, Bernoulli numbers, Lucas numbers, Ramanujan numbers, Euler numbers, Germain primes, Mersenne primes. Now consider a few of the areas in our lives in which numbers are major players—

Bar codes have been invaluable inventions to keep track of merchandise. And of course every bar code has a number attached to it. Now we even find bar codes are going to be used to keep track of things on the molecular level. SurroMed company, based Mountain View CA, has developed a way to tag barcodes to track genes, proteins and other molecules.

  • numbers identifying the passage of time from nanoseconds to millennia.

• the base 60 time on the microwave and clocks.

• economic numbers measuring the status of economies from unemployment numbers to inflationary index to GNP(Gross National Product) & GDP(Gross Domestic Product).

Number systems and symbols used over the centuries												
Hindu-Arabic base 10	0	1	2	3	4	5	6	7	8	9	10	11	12
Babylonian base 60	𐎟	𒁹	𒈫	𒐈	𒐉	𒐊	𒐋	𒐌	𒐍	𒐎	𒌋	𒌋𒁹	𒌋𒈫
Ionic Greek numerals		A	B	Γ	Δ	E	Ϛ	Z	H	Θ	I	IA	IB
Egyptian hieroglyphic numerals base 60		I	II	III	IIII	II III	III III	III IIII	IIII IIII	III III III	∩	∩I	∩II
Chinese script numerals		一	二	三	四	五	六	七	八	九	十	十	十
Hebrew numerals		א	ב	ג	ד	ה	ו	ז	ח	ט	'	יא	יב
Chinese rod numerals base 10		I	II	III	IIII	IIIII	T	TT	TTT	TTTT	⊟	–I	–II
Roman numerals		I	II	III	IV	V	VI	VII	VIII	IX	X	XI	XII
Egyptian Hieratic numerals		I	Ս	III	Ⴑ	﹁	Ʒ	⟨	⟨	⟨	∧	∧I	∧Ս
Mayan numerals base 20	👁	•	••	•••	••••	—	▬	▬▬	▬▬▬	▬▬▬▬	═	═	═
binary numerals base 2	0	1	10	11	100	101	110	111	1000	1001	1010	1011	1100

• numbers which measure temperatures—Celsius, Fahrenheit, and Kelvin for example.

• numbers that make the function of the electronic computer possible — binary, hexadecimal, bauds, ram, rom, bits, bytes, hertz.

• household numbers — microwave, digital, water, electricity, gas, heating, cooking numbers.

Some of the numbers mathematicians use in their work & studies

√8 e 0 -7 4 π googol -2+3i .625 1,238,901

1
3/8
√-1
ϕ
+67
220
5th
𝒳₀
19

RATIONAL   TRANSFINITE
amicable
COMPLEX imaginary prime ordinal algebraic
WHOLE positive
composite
decimals negative perfect odd figurate real
natural INTEGERS
even *fractions* cardinal
IRRATIONAL *counting*

35.7898989... 284 -√4 .333

- earthquake numbers — Richter scale numbers.

- stock market numbers — Dow Jones averages.

- health numbers — grams, blood pressure numbers, blood counts, smog numbers, sound decibels, calories, body temperature, eyesight (20/20), EPA (Environmental Protection Agency) numbers, bone density, age.

- automobile numbers — speed, gas levels, mph.

- educational numbers — test scores, GPA(grade point average), SAT scores.

- sport numbers — batting averages, yard gains/losses, scores, boxing count down, RPI(runs per inning).

• privacy numbers — PIN numbers, social security numbers.

• entertainment numbers — radio bands, TV channels, movie ratings.

• political numbers — vote counts, pole numbers.

• social numbers — crime rates, divorce rate, birthrate, telephone numbers.

• nature's numbers — tide numbers, lunar and solar cycles, astronomical numbers, air pollution numbers, gravity, ice density.

• bar code numbers to identify almost everything we buy.

The list can goes on and on. Numbers are everywhere. Most of the time we don't realize how pervasive they are. When we begin to look, we realize we have indeed been invaded by numbers. Modern society can no longer function without them. Numbers can be our friends or our nemesis, it all depends on our perspective. What isn't in some way connected to numbers?

You've come a long way π

Many of us prefer problems with solutions that are nice "round" answers — no answers with extra baggage as in fractions or decimals. In fact, most of us are perfectly satisfied getting integers. But integers don't necessarily have all the "good" stuff. Take a look at 3.1415926...; 1.4142135...; 2.7182818... What secrets do these "bizarre" numbers hold in their decimals? They hold answers to questions that are either very simple or very complicated. π, for instance comes to mind when considering the size of any circle. Look at the diagonal of any square and √2 =1.4142135... is always connected to it. Glance at how the interest in your saving account is computed— yes, 2.71828... = e is a major player. These "weird " numbers are in our lives explaining and governing its order or disorder.

> The four more widely known irrational numbers are:
>
> π = circumference/diameter
> ≈ 3.1415926
>
> $e = \lim_{n\to\infty}(1+\tfrac{1}{n})^n \approx 2.7182818$
>
> $\sqrt{2} \approx 1.4142135$
>
> $\phi \approx \dfrac{1+\sqrt{5}}{2} \approx 1.6$
> the golden mean

In an arbitrary division problem the remainder is often rounded off. But there is a famous remainder that is especially important. In fact, its impact on mathematics and our lives has been phenomenal, and has spanned millennia. Who would have thought that the problem of dividing the distance around any circle by the length of its diameter would introduce

Some people have been challenged by memorizing 𝜋's digits. Mathematician Simon Plouffe, who memorized the first 4096 of its digits as a teenager, has coauthored with Fabrice Bellard and Jonathan Borwein a fabulous equation that enables you to find any digit of 𝜋 without calculating any previous digits. The catch is that the digit is in hexadecimal (binary) form, and cannot be converted into a base ten digit without knowing all the previous base ten digits. For example, if 9 in binary form is 1001, we see its third digit is 0, but that says nothing about its base ten form without knowing all its digits.

In 1995 Hiroyuki Goto memorized 𝜋 out to 42,195 digits.

one of the most sophisticated mathematical numbers— the famous and infamous 𝜋 — 3 plus a little bit more? Over the centuries its exact value has eluded the most adept mathematicians.

In 1610 Ludolph van Ceulen was the first to use decimals to try to express 𝜋's remainder. Ever since then π's decimals have been teasing, tantalizing and frustrating lay people and mathematicians. There are many anecdotes of people trying to unravel its decimals, find its end, memorize thousands of its digits or look for patterns in its never ending decimals. Even though many things have been discovered about 𝜋, our fascination never seems to cease. Most problems with remainders do not create the uproar that π has. Today, we are no less fascinated than ancient civilizations to discover that the *"magic" circle number* always is the same regardless of the circle's size.[1] The ancient Babylonians and Egyptians discovered approximate values by manually

measuring circles with their respective diameters. In 2000 BC the Babylonians used 3 1/8 for 𝜋, and in 1700 BC the Egyptians had the value 256/81. But the first major breakthrough in calculating 𝜋 is credited to Archimedes (c. 287-212 BC) who

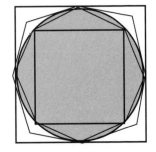

inscribed and circumscribed a circle with two regular polygons. As the number of sides of these polygons were increased, their perimeters came closer and closer to that of the circle's circumference they enclosed. Thereby, Archimedes was able to pinpoint π's value between 3 10/71 and 3 1/7. It was Archimedes' technique that Ceulen used to calculate π's first 35 decimal places in 1610. Ceulen used two squares to enclose a circle. Then he made the squares into two regular octagons, then two 16-gons, etc. To calculate π's 35 decimal places he ending up using regular polygons with 262 sides. Around 1665 Isaac Newton introduced calculus, and used it and his binomial theorem to devise methods involving infinite series to calculate a string of π's decimals. Other mathematicians took up the tools of the calculus of both Newton and Gottfried Leibniz. Working with calculus, Abraham Sharp found π to 71 places in 1699. Seven years later John Machin found π to 100 digits. In the same year, 1706, π was designated for the first time with the Greek symbol π.

Among the many mathematicians searching for π from the 3rd century BC to 1700 are Archimedes, Ptolemy, Ch'ang Hong, Tsu Ch'ung, Aryabhata, Brahmagupta, Fibonacci, François Vieta, Ludolph van Ceulen, Willebrod Snell, Christian Huygens, John Wallis, Muramatsu Shigekiyo, Isaac Newton, James Gregory, Gottfried Wilhelm Leibniz, Abraham Sharp. From the 18th century through the mid 1900s century we have — John Machin, William Jones, Thomas Fantet de Lagny, Takebe Kenko, Leonard Euler, Johann Lambert, Georg Vega, L.K. Schulz von Stassnitzky, John Dase, Tseng Chi-hung, Ferdinand von Lindemann, D.F. Ferguson. Finally, from the mid 1900s to the present mathematicians have programmed computers to calculate π's decimals.

In 1671, James Gregory discovered the series expansion for $\pi/4$. In 1673 it was reinvented by Leibniz, whose name is attached to it today.

William Jones was the first to use the Greek letter π to represent π. It appeared in his book *Synopsis palmariorum matheseos* in 1706.

Until then, it was described as C/d or as an approximate fraction or decimal. In 1767 Swiss mathematician Johann Lambert proved that π was an *irrational number* —confirming that it could not be written as a fraction and its decimal representation would never end or repeat. You would have thought that such a finding would lessen the infatuation for π's decimals. But no, in the 1800s human computers continued to work on calculating out π's decimals and

Transcendental numbers are numbers which cannot be roots (solutions) of polynomial equations with rational coefficients. When π was proven transcendental, this also implied that the *ancient problem of squaring a circle* — (Given a circle with radius r, construct a square of equal area using only a compass and straightedge.) — was impossible.

searching for patterns. In 1873, after 20 years of manual computations, William Shanks calculated π to 707 digits. His record lasted until 1946, when D.F. Ferguson first published a list of the first 620 decimals and discovered an error at Shanks 528th digit.

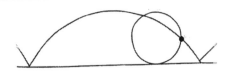

In the 17th century the *cycloid* (the curve traced by a point as a circle rolls along a line) and its properties were studied in depth. The length of one revolution of a cycloid's curve was discovered not to be connected to π, but to be a rational number equal to 4 times the rotating circle's diameter. The area under the cycloid, on the other hand, was indeed connected to π — 3 times the area of the rotating circle, $3\pi r^2$.

Most problems which deal with circles have π connected to them in some way. The centuries are speckled with π problems and endeavors. The ancient Greek problem of trying to square a circle involved π. If it were possible, this would have meant that a square could be found whose area equaled that of a given circle. The search for this problem's solution led to the discovery that π was a *transcendental number,* and in 1882 German mathematician Ferdinand Hanover proved π was a transcendental number. But the fascination for π's decimals did not ebb with this discovery. The first ground breaking computer work on π occurred in 1949 on the ENIAC (Electronic Numerical Integrator and Computer). Mathematicians and researchers George Reitwiesner, John von Newmann, and N.C. Metropolis programmed the ENIAC which calculated π to 2,037 digits in seventy hours. Years later digital computers entered the picture, and π's decimal places took off. Mathematicians came up with quicker formulas and programs for these new powerful computers that processed and crunched the numbers so that by the 1990s the 707 decimal places of 1873

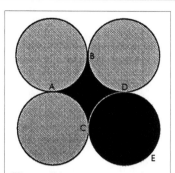

The well known *urn puzzle* made from circles has a solution not involving π. The area of the shaded black urn shape (ABCDE) can be shown to be 4 times the radius squared.

were now in the billions. Over the next 50 years companies, researchers, universities began to use π to make a name for their, computers and themselves. Here we find the famous Chudnovsky brothers[2], David and Gregory, who built and pro-grammed their own computer to carry out their computations, and in 1966 their computer calculated π to 8 billion digits. A year later Yasumasa Kanada and Daisuke Takahashi[3] had a Hitachi SR 2201 calculate π to over 51,539,600,000 digits in a little over 29 hours.

What does one do with all these decimals? They put supercomputers to work by testing their speed and reliability at generating π's decimals. But this motive did not drive mathematicians who lived in BC (**Before Computers**). *In hindsight, it was not the decimals that were important, but the ingenious mathematics that brilliant minds uncovered.* Consider some of the mathematical insights and discoveries the search for π's digits fueled —irra-tional and transcendental numbers— work with infinite sided figures — new trigonometric formulas —infinite series and other calculus methods and techniques —computer technologies and programming techniques. All of these apply to so many of today's varied fields of mathematics. As the authors of *Pi: A Source Book* state *"... to pursue this topic (pi) as it developed throughout the millennia is to follow a thread through the history of mathematics that winds through geometry, analysis and special functions, numerical analysis, algebra, and number theory. It offers a subject which provides mathematicians with examples of many current mathematical techniques as well as a palpable sense of their historical development."*[4]

Will mathematicians ever get enough of π? As Galileo Galilei said *"the universe...is written in the language of mathematics ...without which it is humanly impossible to understand a single word of it...".* What other secrets of the universe does π have hidden in its digits? Perhaps mathematicians will discover these in less time than π has been known.

---

[1] It always takes 3 diameters plus a little bit more to cover the distance around the circle·

[2]Their work with π is used by them in such areas as hypergeometric function identities and Diophantus approximations.

[3]They used an algorithm which was devised by Eugene Salamin in 1976. Salamin later discovered that his algorithm resembled one which Karl Gauss had developed in the 1850s for calculating elliptic integrals.

[4] *Pi: A Source Book* by Len Berggren, Jonathan Borwein, and Peter Borwein, Springer-Verlag, New York, 1997.

icosa shelters
mathematically designed shelters

Ancient polyhedra and Platonic solids inspired Leonardo da Vinci's solid ornaments and Buckminster Fuller's 20th century geodesic domes. Now the ancient icosahedron shape is making its debut in the 21st century in ICOSA shelters designed by Sanford Ponder. Icosa shelters are being used as recreational or work shelters ranging in diameter from 9 to 23 feet. Because of their portability they can easily and effectively be utilized in areas where relief efforts are taking place. The 23 foot shelter kit is shipped unassembled and weighs about 500 pounds. The beauty of these shelters is that they are very sturdy yet light weight. In addition, they can withstand high winds, can be steam cleaned or washed with disinfectant, and safely burned.

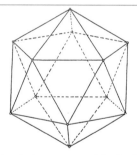

The icosahedron was one of the five Platonic solids which were shown to be the only regular(all faces and sides were the same shape and congruent) polyhedra.

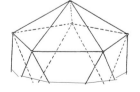

What does an icosa shelter look like when completed? An icosahedron cut in half

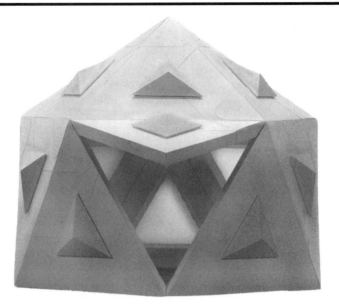

For other models and information on the icosa house see website http://icosavillage.net for further information.

with windows inset within its triangular walls.

Ponder's icosa shelters are formed —

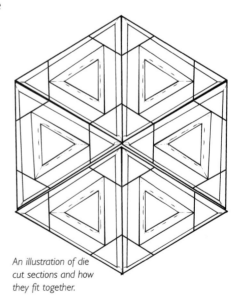

- by using equilateral triangular shaped pieces die cut from 100% recyclable polyolefin material or fiberboard.

- by studying the model of an icosahedron, geodesics and tensegrity.

- by folding the die cut parts as scored, and thereby making a semi-icosahedron-like structure.

An illustration of die cut sections and how they fit together.

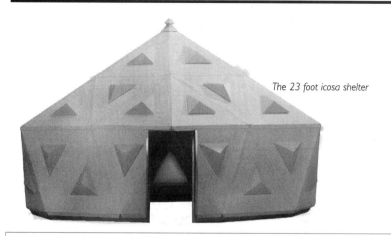

The 23 foot icosa shelter

The *geodesic dome* was one of Buckminster Fuller's famous creations. The dome forms a triangulated surface which encloses space and its shape approaches the curvature of a sphere. The beauty of these structures is that they can enclose far greater volumes using less materials than other architectural shapes. As Fuller writes in his patent application of 1954, *"A good index to the performance of any building frame is the structural weight required to shelter a square foot of floor from the weather. In conventional wall and roof designs the figure is often 2500 kg per square meter. I have discovered how to do the job at around 4 kg per square meter by constructing a frame with a skin of plastic material."*

Tensegrity — *tensional integrity is created by the pulling action of a load where an equilibrium is reached between tension and compression* — it is what gives stability and strength to the geodesic domes.

Geodesic dome shapes also appear on a molecular level. The *buckyball*, also known as the *buckminster-fullerene*, C_{60}, is a chemical polyhedron synthesized in 1990s,. It consists of 60 carbon atoms located at the vertices of a truncated icosahedron.

truncated icosahedron— a buckyball's model

Computers get stressed out too!

lock-up, crash, freeze

Most people at some time in their lives are concerned about aging and the effects of stress on one's physical and mental well being. Well guess what? Computers are not immune to stress and the aging process, either. They age, get stressed, and practice self-healing. Computers can begin to slow down or even stop, often in the middle of executing a command or function. What causes these problems in computers? Sometimes computers are burdened (i.e. stressed) by too much information. Most of the time, this stress is cumulative, similar to that proverbial last straw that put your back out. "Health issues" result either from the invasion of a "disease" (such as a virus or a worm) or simply the accumulation of coding abnormalities inherent in software that slow down, consume memory, or even freeze up the computer. During this process, the microprocessor reaches a stalemate resulting in indecision. It literally does not know what should be

A *crash* is a system failure in your computer which requires you to intervene by closing down your computer or even performing some maintenance on its system before restarting your computer using that system.

A *freeze* or *lock-up* is also a system failure which stops your computer dead in its tracks. Most of the time keystrokes or mouse functions do not work nor can they intervene to revive the computer. Usually there is nothing you can do other than restart your computer.

done next. Software programming errors also put a drain on the computer's memory. This happens to almost all personal computers. Sometimes there are conflicts between various programs or old programs trying to be used on computers with updated newer systems. When this happens the mouse may not respond, the keyboard is dead, the cursor is frozen or begins to incessantly spin. Sometimes you can hit escape or force quit and clear it out, but other times the only thing that works is to turn off your computer. All at the expense of lost unsaved data. Shutting off the computer usually resets things, especially the software culprit. But, some errors remain within the microprocessor. There are various software fixes that give your computer an examination or check-up which uncovers and fixes these coding abnormalities, but one usually does this after the problem occurred.

A *server* is a computer on a network which provides various services to the computer of the network, such as hosting your website, connecting you to other websites, sending your e-mail.

A *mainframe computer* is a large powerful computer which serves as a central system shared by users often over a private network of computers.

Now imagine the stress and aging a server computer

experiences. Servers are not shut down as often as one's personal computer. Unlike PCs, shutting down a server computer to perform examinations means a lot of down time which results in a lot of angry customers and large revenue losses. Computer companies, such as IBM, are developing rejuvenation software — a computer's own health spa treatment. Actually the software is being designed to predict failures before they happen, rather than react to them after the failure occurs. Even as a virus program detects and exterminates a virus before it completely takes over your entire computer system, the rejuvenation program would predict failures before forcing servers to react to a lock-up. It does this by measuring and recording the amount of work a server performs, and with these numbers it quantifies its stress. By analyzing a computer's stress and predicting an impending failure, the software would automatically schedule a convenient down time after transferring tasks and data to other servers and thus taking care of the problems.

What about mainframe computers? Mainframe computers rarely experience such stress since their software and hardware are carefully checked for compatibility and carefully scrutinized for errors.

System and programming errors are all linked to logic — the mathematical way of thinking. But sometimes programmers don't get it right, at least for computers — causing computer "burnout".

mathematics &
the pomegranate
the rhombic dodecahedron & sphere packing

Recipe for rhombic dodecahedron:

• Take two identical cubes.

• Carefully slice one.

• Add it to the other.

You've just cooked up the amazing rhombic dodecahedron. It is not just your everyday polyhedron but one with a colorful history.

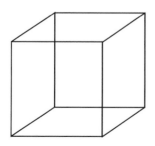 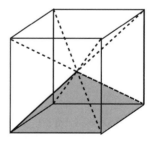

Notice the cube on the right is sliced into 6 congruent square pyramids. One of the pyramids is shaded. If you were to place one pyramid on each of the square faces of the cube on the left, a rhombic dodecahedron is formed. See next diagram.

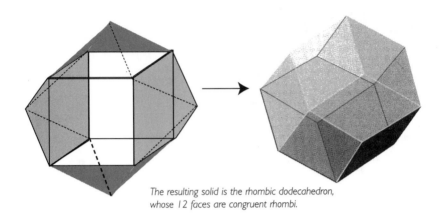

The resulting solid is the rhombic dodecahedron, whose 12 faces are congruent rhombi.

There are only a finite number of shapes that tessellate space, and the rhombic dodecahedron is one. Among others, we find the cube, the rectangular parallelepiped, and the truncated octahedron.

The rhombic dodecahedron's story begins in the late 16th century. Mathematician Thomas Harriot had been asked by Sir Walter Raleigh to come up with a method to easily calculate the number of cannon balls in each stack on his ship. Harriot came up with a simple formula but wondered if the way these cannon balls were stacked (like oranges in a grocery display) was the most efficient manner. Harriot shared the problem with astronomer/mathematician Johannes Kepler (1571-1630), whose curiosity led to the famous *sphere packing problem* which remained unsolved until 1998. The problem came to be known as *Kepler's conjecture.* Kepler felt that the most efficient way to stack identical spheres (i.e. to get the highest density of volume occupied) was a face-centered cubic arrangement. The face-centered cubic arrangement is how produce people have stacked oranges and apples for centuries.

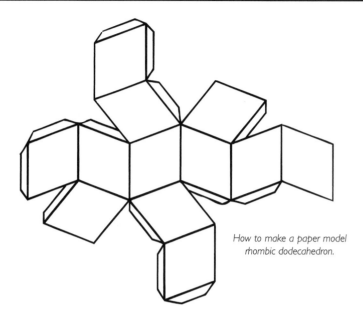

How to make a paper model rhombic dodecahedron.

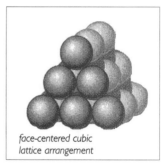

face-centered cubic lattice arrangement

What does this have to do with a rhombic dodecahedron? In the process of reaching his conjecture in 1611, Kepler turned to nature for inspiration. After all, so many natural phenomena illustrate the conservation of resources and energy.[1] Kepler focused his attention on the tiny spherical seeds packed into a pomegranate's shell. He observed that the seeds were initially packed in a *face-centered cubic lattice*. And as they grew, expanded, and filled the interior space, their shapes were transformed into rhombic dodecahedra — hence his discovery of this 3-D tessellating polyhedron. He looked at three ways of arranging identical spheres.

cubic lattice arrangement

Their technical names are the *cubic lattice arrangement, the face-centered cubic lattice,* and the *hexagonal lattice.* The term lattice refers to a grid of points, such as (x,y,z) coordinates of cubic space. In this case, the points of the lattice are the spheres' centers. From here, Kepler computed the density (how much volume of a

hexagonal lattice arrangement

cube is occupied by the spheres) for each of these three arrangements. He determined the cubic lattice gave $\pi/6$ (≈0.5236), the face-center was $\pi/3\sqrt{2}$ (≈0.7404), and the hexagonal gave $\pi/3\sqrt{3}$ (≈0.6046). In addition to these three ways, there are infinitely many other ways to package the spheres. Was Kepler correct in asserting that the faced-center method packed the spheres in the most dense way? [2]

Johann Kepler (1571-1630)

A simple mathematical question evolved into a host of other ideas — from a military problem, to the discovery of the rhombic dodecahedron, to the sphere packing problem. But the story does not end here. The sphere packing problem has moved from 3-dimensional space to higher multi-dimensions and deals with n-dimensional hyperspheres[3] leading mathematicians to major breakthroughs in code theory[4] — the branch of mathematics which deals with transmitting and recovering data and messages over static and noisy channels. Code theory ties into both information storage on compact disks and compressing information for transmission not only via the internet but throughout the universe.

---

[1] Examples — nature's sphere encloses the most volume with the use of the least surface area; the bee's hexagonal honeycomb uses the least material for a given area.

[2] Different stages of the problem were tackled by mathematicians over the centuries. Karl Gauss (1777-1855) made major headway when he proved it was for all packing in which the spheres centers were arranged in a lattice. But what about the non-lattice arrangements? In 1953 Hungarian mathematician Laszlo Toth reduced the problem to solving specific cases requiring an enormous amount of calculations. The final proof with computer calculations was done by Thomas Hales in 1998 and consisted of over 250 pages in addition computer programs and data.

[3]See chapter on *the mathematics of higher dimensions*. A glimpse at how mathematics explores higher dimensions, for more on multi-dimensional spaces and hyperspheres

[4] Code theory is not to be confused with cryptography which is the science of encrypting messages to keep them secret.

the eyes have it

the science of biometrics

I n the past, a calling card was a note or card announcing who you were, and was presented when you called on someone. In today's world, calling cards are one way of communicating by telephone. Once a valid PIN number has been entered, we are able to gain access and charge the call to our account. Today numbers and digital technologies serve as our letters of introduction or keys to our bank accounts, telephone accounts,

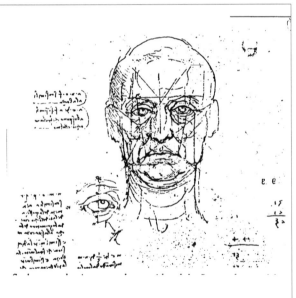

Sculptors and painters can be considered the first analyzers of the structure of the human face and body parts. This sketch by Leonardo da Vinci shows his study of a model's face. The science of biometrics mathematizes one's individual traits for purposes of identification. Today face recognition software is able to analyze and compare a scan with thousands of others in its data base.

computer usage, and even to job sites. Establishing who you are is not as social as it was in the past. Interfacing with another person to identify yourself is not always possible in our digitally connected

> *biometrics—The* science of quantifying or measuring the characteristics of living things

world. Modes of accessibility may ask for fingerprints and voice recognition plus a code number. Whereas in the past a signature sufficed as evidence for an agreement, many states now require a notarized signature and a fingerprint for documentation involving real estate. With the introduction of various electronic forms of acquiring money or information — be they ATMs or smart cards or just to access your e-mail — PIN numbers are required. For a driver's license some states require photographs and fingerprints. In the 1960s, many banks were still entering passbook transactions manually. You could go to your local bank and speak directly to a manager about a loan, and the manager would have the authority to grant or reject your request. They knew who you were. This is no longer a common occurrence. We now can get instant cash in any city at anytime of day with a plastic card at an ATM machine. A plastic card

Wave theory relies on the mathematics of sinusoidal curves — curves whose shapes repeat themselves. These are used to describe waves, in particular sound waves. Since sound has a distinct curve to describe it, each voice can be identified by its own sinus curve and sinus equation with its own particular amplitude and frequency — i.e. pitch & rhythm. Fourier analysis has expanded the ideas of wave theory. Using what are called windows which zero in on a specific minute sound by studying tiny increments with sine wavelets used just to describe that part. When these windows' increments are linked to time intervals, which can also be as small as one wants, they are called Daubchies wavelets (named after the mathematician who developed them, Ingrid Daubchies), and their descriptions become every so accurate.

The *binary number system* is the means by which information is transmitted and communicated to computers. Electrical current exists in only two states *off* and *on*. These two states are described by the binary number system, since this number system relies only on 0s and 1s to write numbers, symbols, or information.

Cryptography, the science of writing codes and ciphers, relies heavily on mathematics of number theory. One area of number theory deals with prime numbers. Encryption programs use enormous prime numbers to secure information that is transmitted electronically.

and a signature also lets us eat out, travel, buy gas, groceries and gifts. The convenience of plastic cards spawned abuses and fraud. We now have an identity crisis. A PIN number and signature are no longer enough protection. Thus, the science of biometrics has developed more precise methods of identification.

Biometrics uses mathematics to identify animate things by quantifying various physical and behavioral characteristics or personal traits. Mathematics and computer science converts these traits to 0s and 1s so they can be stored

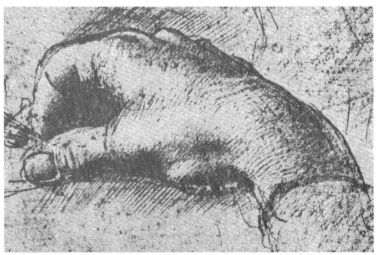

The hand and its geometry has been studied and sketched by many artists from Leonardo da Vinci to Auguste Rodin to M.C. Escher, and today it has become a focus of the of biometrics. Sketch by Leonardo da Vinci.

digitally. Theoretically, each individual's information would be stored in a data base where it can be accessed for comparison whenever access is requested. Scanners, cameras, sound recorders, and computers are forms of the technologies used to measure, translate digitally and store and retrieve the data. The mathematics of *wave theory, binary number systems,* and *cryptography* are but a few of the many areas of mathematics used in biometrics.

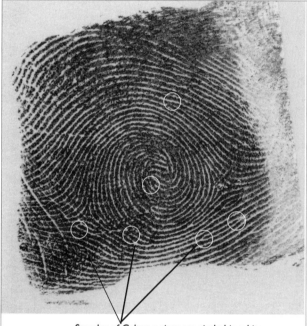

Samples of Galton points are circled in white.

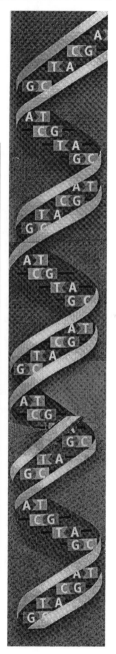

Biometrics is perfecting the art of identification. In the past, latent prints from a crime scene were manually compared by examiners to a database of prints. Each fingerprint has distinctive marks, called Galton points. These include the endpoint of a print's line or the intersection of two or more print lines, as illustrated in the diagram. Nowadays a print's Galton points are transformed into a digital pattern of the print using mathematical equations and coordinates. This information is stored in a computer databank. Computer programs quickly compare prints electronically looking for a match to those stored in a data bank. Since DNA can now be quickly and accurately sequenced, forensic science has added DNA sequencing to its tools. A sequence can be derived from only a microscopic sample in very little time.

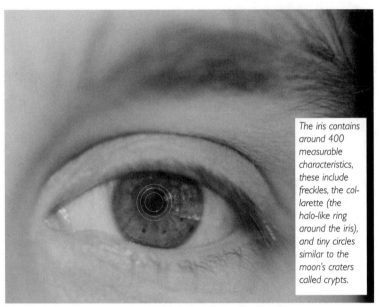

The iris contains around 400 measurable characteristics, these include freckles, the collarette (the halo-like ring around the iris), and tiny circles similar to the moon's craters called crypts.

At an ATM equipped with an iris scanning device, when the customer approaches within 3 feet of the ATM machine, one camera scans the iris of one eye while another takes a black & white photograph. The technology then translates the information into a barcode which is then checked for a match in the existing data base or with the ID card submitted.

Personal traits include fingerprints, hand geometry[1] (which may include palm prints, hand and wrist vein patterns), DNA codes, the physiology of the eye's retina and/or iris, and the face. Voice prints and signatures are considered behavioral traits, because they can be influenced by our emotions. All of these are part of one's persona and have varying degrees of uniqueness and reliability. Without the advent of computers and digital technologies, the collection, converting, transmitting, storing, matching and retrieving of information would require hundreds of hours per person per characteristic. Instead, biometric sensors, such as scanners or digital cameras, convert a physical attribute to a digitized pattern and becomes part of your permanent file in a data bank. All this can be done in a matter of seconds. When a fingerprint is scanned, a digital pattern of its ridges, swirls and indentations is formulated which contains at least 40 measured and recorded features. The chances of two people having identical fingerprints is about one in a billion. On the other hand, eye biometrics have up to 400 measurable features. These include the retina's blood vessel patterns, the many features of the iris, which include pigment granules, fibrous and vascular tissues and subtle movements. A biometric of the eye is done using an infrared light to scan the retina and iris. The results produce a digital barcode of your eye. Thus far the eye biometric is the most reliable detection tool, barring the use of DNA biometrics. Today, the old adage *the eyes are the mirror of the soul* could be extended to *the eyes are the mirror of the soul and hold the keys to private affairs*. Although eye IDs are not yet widely used, the science of quantifying and recording the data is in place. In the wake of September 11, 2001 the use of biometrics is expanding daily. Companies are now making biometric devices available for personal use. In fact, as an individual, you can purchase a fingerprint scanner[2] about the size of a mouse for your computer security. When installed, this device analyzes your fingerprint and allows you to connect or log onto your computer.

Where do we find biometrics being used today? National ID cards encoded with an individual's biometrics are being used in Italy and Hong Kong. Iris scans are used to check workers who commute between Malaysia and Singapore. Israel uses facial recognition biometrics along the Gaza Strip. Frequent air travelers who have their biometrics on file with authorities can bypass customs and immigration checks in the Netherlands and the USA. Mexican workers now crossing the US border use ID cards with finger scan codes for verification. Punch time cards are basically a thing of the past; instead many factories use finger scanning and hand scans readers. The hand's geometry is also used to control access to secured areas for employees at the San Francisco airport, and similarly for travelers at the Tel Aviv airport in Israel. To curb voting fraud, Mexico and Uganda use voter registration cards imprinted with biometric information. Motor vehicle departments across the United States are exploring forming a national data bank with their information and boosted with biometric information. Facial recognitions biometrics has been used at specific events or locales, such as the Super Bowl in Tampa, FL 2001, and at gambling casinos. Among the biometric companies we have InVision, Digital Biometrics, and Drexler Technologies.

Along with the new forms of security offered by biometrics comes a price tag — the sacrifice of privacy. When we order something from a catalog, on line, or even purchase something from a store, we run the risk of having our mailing address shared and sold to other vendors along with our buying habits. Will biometric information — very personal and sensitive information — also be shared[3]? Time will tell.

---

[1] The hand punch terminal is often used by many businesses, and eliminates the use of time cards and badges. A hand is placed into the device which measures the unique size, shape of a hand to verify its identity (no fingerprints or palm prints are used) and records the time the employee begins works.

[2] You simply press your finger on a mini screen (about 1"X2") to analyze and record your print. Once the procedure is followed you no longer need a password to log on. It sells for between $125-$190.

[3] In February of 1998, Senators Steve Peace and Jackie Speier of California presented SB 1622 , which would have required persons, businesses and institutions other than law enforcement to obtain consent before they could collect biometric identification from a person. In addition, it would have prohibited the making of a searchable data base of biometric identifiers, except for law enforcement agencies. It did not prohibit the use of identifiers, nor the collection of identifiers in a searchable data base, merely that any such searchable data base would have to be searchable on some other basis, such as a customer's name, and not on the basis of the biometric identifier. The bill also prohibited selling or sharing of biometric identifiers, except to a law enforcement agency pursuant to a search warrant. The bill provided payment of $25,000 to a person whose biometric identifier is stolen or provided to others in a way other than described above. This bill is currently inactive, and was returned to the Chief Clerk in August 1998. It has not been picked up since then. Since the September 11, 2001 terrorists tragedy, government agencies have been exploring various means to improve security and safety. For example, Senator Dianne Feinstein introduced legislation to create a smart card system for attaining travel visas.

Sol LeWitt

art & mathematics

S ol LeWitt's art, whether intentionally or not, introduces us to a wealth of geometric forms and the mathematics of problems and ideas inherent in permutations and combinations. Consider his work *Incomplete Open Cubes (1974)*. It gives a glimpse at the progression of a mathematical solution, dealing with all the ways edges can imply the formation of a cube. As you walk around his sculpture, you feel you are walking through someone's mind as it considered all the forms possible for rendering an incomplete cube. Along your walk you wonder and ponder at the order behind the incomplete cube formations. You see a row with all the possible ways two sides can be arranged to imply the formation of a cube. Then three sides, then four, all the way until you see the cube formed from its 12 edges. Then your mind might turn to the columns of these rows, and try to analyze and uncover the order in their progression. It is a

The diagram illustrates three progressions of Incomplete Open Cubes

mathematical problem uncovering all the possible incomplete cubes. Sol LeWitt rendered the open cube in three mediums — as a sculpture, as a series of photographs, and as a drawing. The individual photographs of each incomplete cube on a black background are also mesmerizing, as is his drawing. He mentally and manually worked out all the 122 possible shapes of the incomplete cube. As he points out " *The concept would drive the work. I had been using open cubes, but in this work I was interested in ways of not making a cube — all the ways of the cube not being complete.*"[1] Other Sol LeWitt's works such as *Wall Drawing, #601(1989)* are also reminiscent of mathematical themes — a square composed of 25 smaller squares each containing various geometric solids. The *Wall Drawing #34 (1980)* is a mural of mathematical forms in vibrant colors done on a rectangular wall divided by two rows of three squares, each painted with a different geometric 2-D object — these include a circle, a square, a triangle, a rectangle, a trapezoid and a parallelogram. The interiors of these objects are composed of vertical lines while their exteriors are made using horizontal lines. As Sol LeWitt points out *"I don't have a mathematical background. I think of the idea of a system as just another way of making art. Geometry is just another thing out in the world that can be used as art, like trees or toes. … I wanted to make it (the art object) dynamic and to use form as a carrier of content.*"[2]

Many of his wall murals are composed of bold bright colorful geometric objects, such as *Arcs in Four Directions (1999)*, and his sculptures are usually composed of 3-D geometric forms. As he points out, *"… I wanted to do this with the logic of using simple forms.*"[3] When one sees his works, one does not always realize the complexity of the ideas behind them because those ideas are often the essence of a mathematical concept rather than all the concept's ramifications.

---

[1] Page 30, *A Conversation with Sol LeWitt* by Gary Garrels, OPEN, SFMOMA, San Francisco, 2000.

[2] Ibid, page 31.

[3] Page 31.

Virtual Reality is for real

mathematics plays with your mind

Ever imagine walking on the Earth during prehistoric times or participating as an astronaut in a shuttle launch? Our conscious and subconscious imaginations can create forms of VR (virtual reality) without technology. Forms of VR are always creeping unobtrusively into various aspects of our everyday lives. Many of us have experienced the virtual operator, who asks questions which you answer by replying orally rather than punching numbers on your phone pad. Some of us have rented headsets for a virtual guide to a museum's art exhibit. A novel can take us on a journey, solve a mystery, allow us to meet characters and be voyeurs to the intimacies of their thoughts and lives. A dream or nightmare merges our conscious and subconscious worlds in another world. The armchair traveler experiences a trip vicariously, while the silver screen transports us to different times and places.

In VR, as in many dreams, you are not only an observer but a participant. You actually interact in the world you have entered. You might find yourself in Meso-America during the time of the Maya actively involved in the construction of pyramids at Uxmal. Over the centuries scientists have explored various kinds of artificial realities. In the 1500s many used the camera obscura to view the outside world as projected in a box or darkened

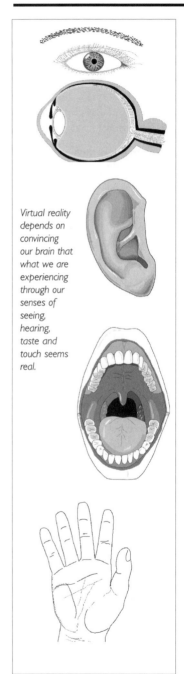

Virtual reality depends on convincing our brain that what we are experiencing through our senses of seeing, hearing, taste and touch seems real.

room. In the 1850s physicists and psychologists explored how our minds were deceived by optical illusions. They studied the physical structure and functioning of the eyes, the optic nerves and the brain in order to explain how optical illusions tricked our minds into believing that these illusions were indeed reality. But virtual reality, as we know it today, was unheard before cyberspace was invented in 20th century. With the advent of the modern digital computer a new type of artificial world evolved. "The computer with its ability to manage enormous amounts of data and to simulate reality, provides a new window…to see reality differently simply because the computer produces knowledge differently from traditional analytic instruments. It provides a different angle on reality."[1]

Originally computers[2] were designed to help relieve humans of drudge work involved in number crunching and storing data. How did we get from a machine which worked on data and spewed reams and reams of raw results to machines which almost instantaneously display solutions? In early computers, information and programs were inputted using stacks of punched cards, and results appeared on reams of computer paper. Eventually the punched cards were replaced by a keyboard that connected directly to the computer, and the computer display made its debut. Advances in radar screens led to computer monitors which could display results almost as instantaneously as the data or instructions were inputted. Scientists, engineers and inventors such as Douglas Engelbart, Ivan Sutherland and J.L. Licklider were among the pioneers whose visions and inventions made these innovations possible. An electrical engineer and a former radar technician, Engelbart first came up with the idea of connecting the computer to a screen which could be used both for seeing data inputted and viewing outputs. In 1962, Sutherland invented the light pen, a device in the shape of a pen which made it possible to draw on the computer monitor via his other innovation the Sketchpad. In the 1970s Sutherland developed the first head-mounted display monitor. At about the same time, Engelbart devised a pointing instrument which allowed text to be manipulated directly on the monitor. These devices were precursors to the mouse and other input devices. With the advent of the integrated circuits and the microprocessors in 1974, computer sizes were scaled down even as their power and memory increased, eventually making it possible for scientists and other interested individuals to own their own computers. The computer, its

Altered reality (AR) is closely connected to VR. They both use similar computer technology, head and body gear, but AR has an entirely different focus. Instead of working in a totally virtual world, which has been simulated using mathematical modeling and high tech graphics, AR takes place in a real environment. This setting is supplemented with virtual information. Compare a VR chemistry laboratory with an AR one. The VR lab is totally designed using simulations. The chemist wears special display glasses or headgear to view the lab and carry out imaginary 3-D experiments. Everything in the lab is digitally simulated except the chemist. On the other hand, in an AR lab an actual chemistry lab is used as the set. Again the chemist dons head gear, but the picture viewed is the actual lab. Virtual test tubes may be used in the lab to carry out experiments. The lab is real, the chemist is real, just some of the lab's components are virtual.

AR requires less computer power and memory to create its virtual parts in its restricted existing environment. VR, on the other hand, has no restrictions on environment but is governed by the computer's capabilities, programs, memory and gear. In the not too distant future, AR may be used:

—to assist a mechanic servicing engines. The mechanic will see schematics along side the engine through the enhanced glasses he or she is wearing.

—to help a surgeon. A scanned image of a patient's organs can be made to appear on the part of the patient's body undergoing the

monitor and the keyboard began to be considered as a single unit. Computers were no longer just tools of government agencies, large corporations, but available to individuals. These innovations laid the foundations for cyberspace and VR was born. New advanced software, computer languages, and graphic programs came on the scene and seduced not only scientists, government agencies, and businesses with what the computer could be instructed to perform, but the general public as well. This amazing tool eventually led to a new form of entertainment — virtual games.

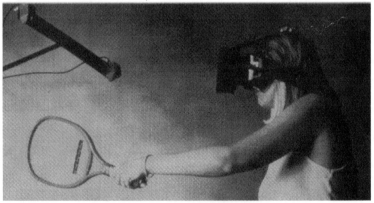

Example of early type headgear used to create a virtual racquetball court. Photograph courtesy of Autodesk, Inc. Sausalito, CA.

surgery. Or a physician may don hear gear to view ultra sound images as the test is being performed on the patient. (This procedure is presently used at the

VR and CS(cyberspace) have often been used interchangeably. Yet VR refers to a perception, while CS refers to a place where the perceptions occur. Neither can exist without the modern computer and

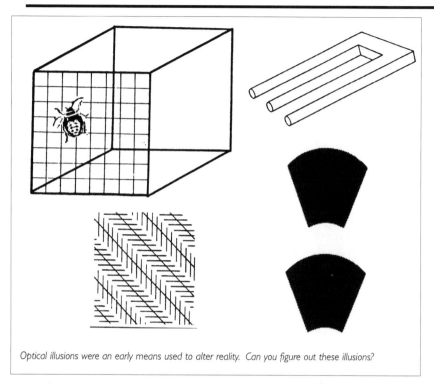

Optical illusions were an early means used to alter reality. Can you figure out these illusions?

mathematics, for computer science and mathematics are the body and brains of VR and CS. VR creates imaginary worlds made possible by digital technologies; worlds composed totally of strings of 0s and 1s rather than actual matter. As Sutherland said in *The Ultimate Display* 1965, "A display connected to a digital computer gives us a chance to gain familiarity with concepts not realizable in the physical world. It is a looking glass into a mathematical wonderland." All is made possible in cyberspace by using computer technologies, mathematical

University of North Carolina at Chapel Hill.)

—to enhance a tourist exploring an archeological site. A program explaining history, virtual images of past, etc. can be activated as the tourist make his/her way around.

— an audience can listen and experience a lecturer, as a 3-dimensional hologram projection, while the speaker is actually making a presentation in another location.

modeling and sophisticated gear[3]. Mathematics defines space as the set of all points. Extending this idea, cyberspace can be considered the set of all points existing within the realm of a computer or a network of computers. Extend the cyberspace of one computer by connecting it to millions of others around the world of the Internet, and cyberspace and our real world appear to become parallel universes. Its in the parallel universes of this global network that new virtual arenas have cropped up in the last decade. Everything that exists in the real world eventually has a cyber equivalent. One can buy a car using a virtual broker, take a virtual tour of a house you may rent from a virtual real estate broker, buy plane tickets from a virtual travel agent, deposit money in a virtual bank, trade stock via a virtual broker, take a course through a virtual class, be an astronomer in a virtual observatory, join a virtual chat room or club. The list is seems endless. While VR can be a form of entertainment and escapism, it is much more. It has sometimes been referred to as a very sophisticated form of optical illusions, but it is much more than that as well. It also can open new 3-dimensional worlds — worlds where we learn, experiment and do research. Its only limitations are the imaginations of those who utilize the mathematics, mathematical modeling, simulations, computers, and hardware necessary to formulate these virtual worlds.

---

[1]Heinz Pagels, *The Dream of Reason,* 1988. *Virtual Reality* , Howard Reingold, Summit Books, New York, 1991.

[2] Over a century ago the term "computer" referred to a person who did calculation, rather than a device.

[3]These include eye gear, data gloves, data body suit.

not your
everyday models
math modeling

Over the centuries mathematicians have devised mathematical models to help solve problems. The ancient Greeks used geometric figures to visualize problems. With the advent of the Cartesian coordinate system, relationships and equations came to view in two-dimensional planes or three dimensional space. Although mathematicians from the past used

> *mathematical modeling* — a mathematical system using mathematical tools such as graphs, equations, computers, and a host of available mathematical fields and theories to describe, understand, and predict physical phenomena.

algebra, calculus and geometry to describe, explain, and predict such things as the motion of falling objects, gravity, fluid dynamics, waves, and the movement of celestial bodies, it is the

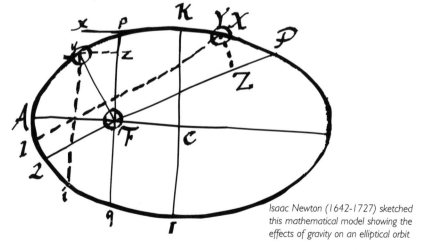

Isaac Newton (1642-1727) sketched this mathematical model showing the effects of gravity on an elliptical orbit

modern computer that has taken mathematical modeling to levels unimaginable by early mathematicians. The computer and mathematics have become inseparable and indispensable tools for analyzing, visualizing and predicting such things as atoms and particles in motion, effects of acid rain, viewing the

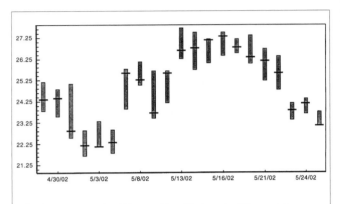

Graphs are one way in which a stock's activity is tracked. The computer can instantaneously extend the stock's activity over a year or over a week or day.

functions of human organs, computer simulations of an earthquake fault zone, and even programs which predict how a child will look at a later age. This dynamic duo has affected the entire spectrum of diverse and apparently unrelated fields — from music to medicine to the stock market to your daily weather forecasting to your movie experience to ecosystems. *How is this possible?*

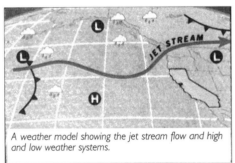

A weather model showing the jet stream flow and high and low weather systems.

Mathematical modeling involves using mathematical descriptions. These mathematical sentences are programmed into the computer using algorithms. An algorithm is just the steps for doing something, i.e. a systematic process of steps which, when followed, completes a task. The results may be displayed in the form of graphs, pictures, 3-D models, animations, etc. For example, the steps of solving an equation or putting on a pair of shoes and socks are examples of algorithms. Algorithms instruct the computer to go about transforming the data inputted in the mathematical model and produce the result(s) (output data) that describe the particular phenomenon being studied or explored. For example, a weather model may use the input data of El Niño Gulf Stream temperatures, global ocean temperature changes, and global satellite data to formulate long or short range forecasts. The model analyzes and presents the results in an easy to interpret format. On the other hand, mathematical modeling has been used extensively in cinematography to produce virtual sets and special effects in such movies as *Star Trek, Titanic, Mighty Joe Young, Matrix* and *Star Wars*. Fractal geometry was used to create imaginary yet realistic sets.

The math is in the bubbles. Yes, even Guinness beer uses m a t h e m a t i c a l modeling to explain the motions of the bubbles in its special brew. For details see *The Physics of Fizz* by Peter Weiss, Science News, May 6, 2000.

One important feature about mathematical modeling is that it gives scientists insights into systems too complex to be studied by only using past methods of linear algebra and calculus. Although these branches of mathematics are powerful and work well with things that undergo constant non-erratic changes, most natural phenomena are too complicated and experience occasional chaotic situations to be adequately described by these tools. The output of such phenomena as weather, disease epidemics, economics, immune systems, and environmental changes may not produce the same output when inputting the same set of data. These mathematical models must be able to adapt and factor in possible interactions and minute nearly imperceptible changes. It is here complexity theory, statistics, fuzzy logic, and probability may enter the picture. Fractal geometry plays an important role in models

Rather than designing a model depicting the bubonic plague as a human disease, Matthew J. Keeling and Chris A Gilligan from Cambridge University in England designed computer modeling following rats that infected people. Their computer modeling suggests that the bubonic plague was never used to describe acid rain and its impact, the ups and downs of Wall Street's stock market, or the realistic scenery in virtual cinematography.

Computers and their models have now become laboratories for many scientists. *Experimental simulations* are a very important part of biology,

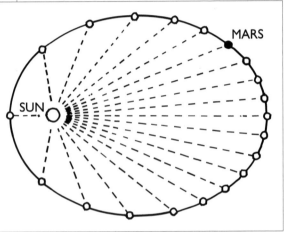

A rendition of Johannes Kepler's (1571-1630) drawing showing the elliptical orbit of Mars. The pie shapes emanating from the Sun as the focus of the ellipse indicate regions of equal areas. He also indicated that the intervals of times between the arc designated by two dots were equal.. Kepler made one of his most important discoveries when he used Tycho Brahe's (1546-1601) observations of Mar's movements. He found that an ellipse rather than a circle fit the observations and his calculations contrary to the then accepted circular orbit belief.

actually eliminated as witness the 1990s outbreaks in Africa and India. Historical data covering the time from the first outbreak in the 1300s substantiates their model, which not only considers the rat population but many subpopulations of rats which may have become immune to the disease. If the subpopulations are not immune and the disease eliminates them, then it can spread to humans. Indicating that killing off the rat population is not the solution since the fleas carrying chemistry, physics— in fact, any science. For example, if scientists need to test nuclear weapons without violating any test ban treaties, without creating nuclear fallout, and without incurring high costs, they can simulate the explosion and its impact with the computer. A molecular chemist wanting to visualize the combining of certain molecules can enter the microscopic world of atoms and molecules through virtual reality programs and software. Mathematical models have also helped environmental scientists and engineers in biorestoration by virtually testing and developing clean-up strategies for polluted sites. Such models must take into consideration interactions between a variety of things such as organisms used to digest pollutants, ground water, various contaminants, and nutrients.

The ramifications and applications of mathematical models may not be readily known when they are being developed. Witness fractals at the turn of the century, or John Forbes Nash equilibrium concept in relation to game theory. These models' applications took years to evolve. On the other hand mathematicians Leszek

the disease would then look for a new host, such as people.

In 2000, researchers at Lawrence Livermore Labs in California and Los Alamos Labs in New Mexico, working independently, developed advanced computer models to show *experimental simulations* of 3-dimensional unfolding of an thermonuclear explosion. This type of modeling eliminates the need to actually detonating nuclear bombs in order to sustain the nation's stockpile.

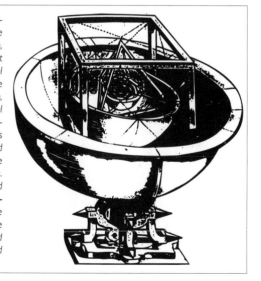

*Kepler devised this "mystical" mathematical model of our solar system. At that time only 6 planets were known, Jupiter, Saturn, Mars, Earth, Venus and Mercury. He felt the uniformity and beauty of this model held a clue to the structure of the universe. At the center he placed the Sun, which makes him the earliest professional astronomer to declare Copernicus' heliocentric that did away with the Earth as the center of the universe. Kepler found that the five Platonic solids separated the six planet and accommodated their orbits. Furthest from the Sun, the **cube** was used to separate Saturn and Jupiter, the **tetrahedron** separated Jupiter and Mars, the **dodecahedron** for Mars and Earth, the **octahedron** for Earth and Venus, and finally the **icosahedron** for Venus and Mercury.*

What does a whip have to do with mathematical modeling?

Scientists had always felt that the crack of a whip was caused by the tip of the whip breaking the sound barrier. But, when scientists at the Ernst Mach Institute in Germany studied this phenomenon using photographs to capture the movement and shock waves, they discovered the shock waves were not created until the tip's speed was twice the speed of sound. Using a mathematical model mathematicians, Alain Goriely and Tyler McMillen of University of Arizona, have theorized that a wave or loop moving along the whip produces the sound by breaking the sound barrier and creating a sonic boom. They modeled the loop's curvature, speed and tension and demonstrated that when the loop reached the speed of sound, the tip of the whip moves at twice as fast creating a mini sonic boom.

Demkowicz and teams from the University of Texas and Texas Institute of Computational and Applied Mathematics are using mathematical modeling to visualize acoustic pressure created by sound waves in the ear canal and head. Their model is being used to improve hearing aids and devices which help the wearer distinguish from which direction the sound is emanating. Demokowicz is also studying the rates of absorption of electromagnetic waves in the head in hopes of shedding light on the health risks of cellular phones.

The amount of number crunching and storage needs created by sophisticated and viable mathematical models is enormous. Even with today's supercomputers able to perform trillions of operations per second, the information processed and produced by these models taxes and overburdens even these most sophisticated computer systems. The demands for bigger, faster computers seem never ending. The more advanced computers become, the more complex will be the problems they will tackle. The 21st century will be a dynamic place for amazing mathematical modeling and problem solving by traditional digital super computers and teams of parallel computers globally networked. It will also witness the emergence of viable quantum, optical and molecular computers each suited to special tasks.

Marcel Duchamp & the 4th dimension

During the mid and late 1800s physicists and psychologists explored how visual reality was altered by optical illusions. Nearly 200 scientific papers were written on the subject. This was also a time when mathematicians were exploring *non-Euclidean geometry* and *n-dimensional space*—mathematical ideas which also altered existing realities. Likewise, books/papers were written which centered around higher dimensions. Edwin Abbott's *Flatland*, published in 1884, has become a classic on the subject of dimensions, and H.G. Wells' *The Time Machine* continues to captivate the public. So it is not surprising to see how artists such as Pablo Picasso, Georges Braque, Marcel Duchamp, Fernand Léger were influenced by the new intriguing ideas of non-Euclidean geometries, Einstein four-dimensional space of length, width, height and time, and optical illusions. In fact the 20th century ushered in bold new forms of art. Art that did not focus on reproducing objects and scenes, but rather on conveying ideas.

The *field of optical illusion* was essentially launched by Johann Zollner (1834-1882) when he stumbled upon a piece of fabric with a design similar to the the one below.

With this illusion began the intense study of optical illusions by scientists such as Herman von Helmholtz, Ewald Hering, Johannes Muller, Albert Oppel, Wilhelm Wundt, Joanne Zollner.

Nude Descending a Staircase by Marcel Duchamp. (1912)
Philadelphia Museum of Art: Louise and Walter Arensberg Collection.

Duchamp was especially interested and intrigued by mathematical concepts saying he "wanted to put painting once again at the service of the mind"[1] rather than just relying on the eyes' perception of reality. In addition, he points out in Pierre Cabanne's *Dialogues* that "The *Large Glass* constitutes a rehabilitation of perspective...For me perspective became absolutely scientific ...a mathematical, scientific perspective...based... on dimensions."[2] He goes on to explain how he used the 4th-dimension in *Large Glass*, when he says "I thought that, by simple intellectual analogy, the fourth dimension could project an object of three dimensions, or, to put it another way, any three dimensional object, which we see dispassionately, is a projection of something four-dimensional, something we're not familiar with."[2].

Duchamp most likely relied on the mathematical writings of French mathematicians Henri Poincaré and Esprit Pascal Jouffret and on salon discussions[3] on these themes for information on the non-Euclidean works of mathematicians Georg Riemann, Nikolai Lobachevsky, and János Bolyai. Duchamp's works in themselves illustrate his study, interpretations and incorporation of these new mathematical concepts. In *Nude Descending a Staircase* the eye is almost at a loss at where to begin to view the work. All sequences of the figure moving down the staircase occur simultaneously, and all facets of the figure are also revealed

Henri Poincaré designed this abstract to represent a hyperbolic world bounded by a circle. The sizes of the inhabitants change in relation to their distance from the circle's center. As the inhabitants approach the center they grow, and as they move away from the center they shrink. Thus they will never reach the boundary, and for all purposes their world appears infinite to them.

Non-Euclidean geometry evolved as mathematicians began to question whether Euclid's Fifth Postulate *(Through any point not on a given line there is only one line parallel to the given line.)* was indeed a postulate and not a theorem. After many centuries and hundreds of attempts to prove it as a theorem, other geometries were discovered which formulated different versions of this postulate and which could only hold true in different non-Euclidean worlds. In 1854 Georg Riemann devised a new interpretation of the Fifth Postulate (*Through any point not on a given lines there are no lines which can be drawn parallel to a given line.)* in which a strange new geometry evolved called *elliptic geometry*. The universe of this geometry consisted of a sphere, and on this sphere's great circles were its lines which always intersected in two points. In other words, there were no parallel lines in this geometry. Here the three angles of a triangle always totaled more than 180°, and as the triangle's area increased the sum of its angles also increased. Among other non-Euclidean geometries we have *hyperbolic geometry, fractal geometry, topology* — all with their own unique mathematical system.

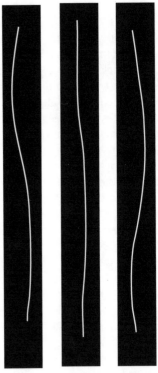

A rendition of a portion of Marcel Duchamp **Three Standard Stoppages,** 1913-1914. It consists of threads varnished onto canvas strips. The complete work is on display at The Museum of Modern Art, NY, Katherine S. Dreier Bequest.

at once from all dimensions. Duchamp describes his work *The Bride Stripped Bare by Her Bachelors, Even* (also called the *Large Glass)* in his publication the *Green Box,* referring to it in terms of various dimensions. In his publication *A l'infinitif* he discusses four-dimensional perspective. His work *Three Standard Stoppages* deals with the movement of an essentially two dimensional object — a piece of thread through space. Beginning with a straight line (i.e. a straight thread), he drops it and lets it land on a canvas. He did this with three different threads, and each landed differently, demonstrating that a geometric shape's form does not remain static as it passes through higher dimensions. As mathematicians pointed out, multi-dimensional space creates curvature on an object. In hyperbolic geometry a triangle's three angles are less than 180° making its sides convex, while in elliptic geometry the triangle's three angles total more than 180° making its sides concave.

Even Duchamp's *ready-mades* — a term he coined to indicate that an everyday object became a work of art when an artist recognized it as such — were connected to mathematics. Some[4] interpret ready-mades as not totally ready pieces of art, but contend that Duchamp actually manipulated them to create various perceptions and perspectives when he photographed them. For example, his photograph of a four hook coat rack shows each hook from a different perspective, which meant the hooks had to be bent or the photograph altered. In this way he

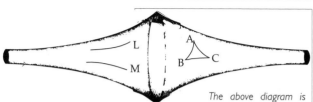

was able to transform a ready-made two-dimensional photograph into a three-dimensional illusion.

It is not surprising that a diverse group of people were intrigued with higher dimensions. In addition to the scientists and the others mentioned above we also find Salvador Dali, Charles Hinton, Oscar Wilde, Dostoyesvsky, Diego Rivera. Yet today the mystique of multi-dimensional worlds has not ebbed. Mathematicians and physicists are exploring higher dimensions in such areas and problems as string theory and sphere packing. Artists such as Tony Robbin seek to catch and create a glimpse of the fourth dimension.

---

[1]Quotes from secondary source, *The Fourth Dimension and Non-Euclidean Geometry in Modern Art*, by Linda Dalrymple Henderson, Princeton University Press, Princeton, NJ, 1983.

[2]Ibid. Footnote 1.

[3]Gertrude Stein's salon on Rue de Fleurus held mathematical discussions at which mathematics teacher Maurice Princet talked about multi-dimensions and the dimension of time.

[4]According the founders, Rhonda Roland Shearer and Stephen Jay Gould of the Art Science Research Lab in NY, and other colleagues.

The above diagram is called a pseudosphere, and is used for a model of hyperbolic geometry devised by Nikolai Lobachevsky and János Bolyai in 1829. In this geometry lines L and M are not parallel in the same way we expect lines o be parallel in Euclidean geometry. They do not remain equidistant throughout. The distance between hyperbolic parallel lines diminishes in one direction and increases in the other direction. These hyperbolic parallel lines are actually asymptotes—curves that continually approach a line but never touch it. On the hyperbolic pseudosphere universe lines are actually curves. Notice the shape of a hyperbolic triangle's sides are convex, making the sum of its angles less than 180°. Also notice as a triangle's area increases the sum of the three angles decreases.

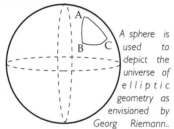

A sphere is used to depict the universe of elliptic geometry as envisioned by Georg Riemann.. Here lines are great circles of the sphere of which any two always intersect in two points. Notice that the sides of an elliptic triangle are concave making its three angles total more than 180°. As the triangle's area increases the sum of its three angles also increase.

There's no business like math business
mathematics & entertainment

Mathematics has been very evident in the entertainment arts lately. Consider these recent works which became Broadway/Hollywood hits: *Arcadia, Goodwill Hunting, Picasso at Lapin Agile, Proof, A Beautiful Mind, Windtalkers,* and *Enigma.* Most of the time when one hears that mathematics plays a prominent part in a movie we think it must deal with such science fiction works as *Contact* based on Isaac Asimov's book of the same name, or the *Star Trek* series. But the above pieces prove otherwise.

In the early 1990s, theater goers were taken by storm by Tom Stoppard's play, *Arcadia* — a story set in two time periods which were related by the evolution of mathematical ideas. For example, Stoppard links elements of Euclidean geometry, and fractal geometry when his heroine, Thomasina, says

> *"...if there is an equation for a curve like a bell, there must be an equation for one like a bluebell, and if a bluebell, why not a rose? Do we believe nature is written in numbers?...Then why do equations only describe the shapes of manufacture?...Armed thus, God could only make a cabinet....We must work outward from the middle of the maze. We sill start with something simple. I will plot this leaf and deduce its equation."[1]*

In addition, the mathematics in *Arcadia* also deals with calculus, chaos, complexity and iterations. Stoppard even mentions Fermat's Last Theorem. These concepts are skillfully and clearly woven into the dialogue and the development of the plot and characters. Here is how one of his characters describes the chaos theory of a drip:

> *"...We can't even predict the next drip from a dripping tap when it gets irregular. Each drip sets up the conditions for the next, the smallest variation blows prediction apart..."* 2

The character Hannah, introduces the use of iterations in chaos theory, when she asks: *"What I don't understand is ...why nobody did this feedback thing before—it's not like relativity, you don't have to be Einstein.*[1]

The mathematician Valentine replies: *"You couldn't see to look before. The electronic calculator was what the telescope was to Galileo."*[4] This amazing play gave the public a painless glimpse at mathematical ideas and an introduction to fractals geometry.

In *Good Will Hunting* we met an unknown math genius played by Matt Damon, who introduced laypeople to the world of mathematical thought.

Then in the mid-1990s *Picasso at the Lapin Agile* by Steve Martin brought the turn of the century worlds of the young Pablo Picasso and Albert Einstein together by having them meet each other in a bar. The snappy dialogue blends the worlds of art and science. For example,

> Einstein: *"I hate to tell you this, but the idea of a triangle with four points will not fly. A triangle with four points is what Euclid rides into hell."* [5]

> Picasso: *"I think in the moment of pencil and paper, the future is mapped out in the fact of the person drawn. Imagine that the*

In the 1600s Pierre de Fermat wrote his famous *Fermat's Last Theorem (FLT)* in the margin of one of his books. *If n is a natural number greater than 2, there are no positive whole numbers x, y, and z such that $x^n + y^n = z^n$.* He claimed he had proven it, but the space of the margin was too small to write in his proof. Leonhard Euler(1707-1783) was the first to prove any part of FLT. He showed that it was true for n=3, but that was 75 years after Fermat presented his problem. This teaser became a mathematical challenge, and the proof or disproof had eluded mathematicians. Sophie Germain (1776-1831) first explored it with a fresh approach. Instead of trying to prove it for a specific value of n, she tackled a class of particular prime numbers, which came to be known as the Sophie Germain primes. A prime number, p, is a Germain prime if 2p+1 is also a prime number. In other words, each Germain prime has to be able to produce another prime when 1 is added to it's doubled. For example 5 is a Germain prime because 1+two5s is prime, namely 11. But, 7 is not a Germain prime because 1+two7s is 15, a non prime.

pencil is pushed hard enough, and the lead goes through the paper into another dimension." [6]

Martin's fictional account captivated the public's interest.

In 2001 the play *Proof* by David Auburn captured the Pulitzer Prize, the Tony Award, and the Drama Desk Award. It continues to play to sold out audiences. It deals with the touchy subjects of how mathematicians burn out at an early age, how mathematicians can become totally consumed by their work, and how women mathematicians cope in a male dominated field. Auburn shows how mathematics crosses genders with a skillfully crafted short dialog introducing the life of Sophie Germain and her work with primes. *Proof* calls attention to Germain primes with the following dialogue:

Hal: Germain Primes.
Catherine: *Right.*
Hal: *They're famous. Double them and add one, and you get another prime. Like two. Two is prime, double plus one is five: also prime.*
Catherine: *Right.*
 Or $92,305 \times 2^{16,998} + 1$.
Hal: *(Startled) Right.*
Catherine: *That's the biggest one. The biggest one known...*

Then the playwright brought the play full circle to the present by focusing on the life and struggles of the heroine in the male dominated world of mathematics. Auburn's play was especially popular because it dealt with relationships (between siblings, parent and children, and mathematicians), the quest for fame, and the thin line between genius and madness. All of these are interwoven in a type of mystery peppered with elements of humor.

Consider the movie *A Beautiful Mind*, based on a novel by Sylvia Nasar, about the mathematician John Nash who won a Nobel Prize in Economics for his work on game theory. Starring Oscar winner Russell Crowe, this Hollywoodized movie deals with the life and work of a brilliant mathematician suffering from paranoid schizophrenia. In addition, in this movie, we find supporting roles played by codes, formulas, logic games.

Yet there are other ways mathematics affects the entertainment industry. Today's flashy films with all their special effects would not be possible without the breakthroughs in computer science that mathematicians have forged. With the 20th century evolution and development of dimensional numbers, fractal geometry, and special computer languages the phenomenal cinematographic scenes viewed in such movies as *Matrix*, *Lord of the Rings, Star Wars, Star Trek* and

Germain's work with this group of primes proved to be an invaluable step in the evolution of the proof of FLT by Andrew Wiles in 1993 Wiles presented his initial version in 1993 at conference in Cambridge, and the final revised/corrected version, which was done in collaboration with Richard L. Taylor, was published in the *Annals of Mathematics* May 1995.

Although *quaternions* were discovered in 1843 by William Hamilton, it was not until the introduction of present day computers that they have come into their own. To mathematically understand how movement is produced consider two points and their coordinates. From Euclidean geometry we know that two points determine a line. The equation can be written in the form $y=mx+b$ (where m is the line's slope and b its y-intercept). As these two points move about on a computer monitor an algorithm can be written using the various equations generated as the coordinates of the points change with their movement, giving a mathematical way of describing a moving object. Here is where quaternions enter the picture. These 4-dimensional numbers provide an excellent way to describe

animated movies would not have been possible or as realistic. Computer graphics are at the heart of digital cinematography. We find the mathematics of fractals creating incredible scenes. In fact, many of these movie sets are being created on computer monitors with the the

actors' roles inserted later. At work here is the mathematics of dimensional numbers, namely 4-dimensional numbers known as quaternions, which make it feasible for the computer monitor to display 3-dimensional moving subjects. Mathematics not only stars in these movies, but also has a prominent part behind the scenes.

rotations in space. A quaternion has a 1-dimensional real part, an imaginary part, and a 3-dimensional vector. Its general form is: $q=a+bi+cj+dk$.

A *quaternion*'s 1-dimensional real part is [a], called a scalar. Its imaginary part is [bi], and its 3-dimensional vector is [cj+dk].) Quaternions have proven invaluable in computer graphics. They make it possible to describe the dynamics of motion in 3-dimensional space. Quaternions play an essential part in software requiring 3-dimensional movement. In fact, quaternions are used to help guide, navigate and fly the Space Shuttle.

---

[1]Page 37 of Arcadia by Tom Stoppard, Faber and Faber, London, 1993.)

[2]Page 48 of Arcadia by Tom Stoppard, Faber and Faber, London, 1993.

[3]Page 51 of Arcadia by Tom Stoppard, Faber and Faber, London, 1993.

[4]Page 51 of Arcadia by Tom Stoppard, Faber and Faber, London, 1993.

[5]Page 49 of Picasso at the Lapin Agile by Steve Martin, Grove Press, New York, 1st edition, 1996.

[6]Page 55 of Picasso at the Lapin Agile by Steve Martin, Grove Press, New York, 1st edition,

Do computers have their limits?

small, smaller, smallest

Ever notice it never seems to fail that just when you've bought the latest state of the art computer it becomes "obsolete"? Why? *Moore's Law* — the number of transistors (or switches) on an individual silicon chip doubles about every 18 months. The more transistors on a chip the more memory and power your computer has. But isn't there a limit? At Moore's exponential rate it would appear that there's only so much room on a chip for just so many transistors. In fact, the current estimate is that by 2010, the size of transistors and the chip will have been shrunk as far as possible —approaching atomic scale — before having to contend with the laws of quantum physics. Chips will have reached their capacity. Will computer power have reached its limit? Will Moore's Law come to a halt? In the past, scientific discoveries and innovations foiled other such predictions. At one point, there was a 1 micron (1/1000mm) limit— predicting that the elements of a circuit could not be made smaller than 1 micron because it was the minimum size of the diameter of the laser beams being used to manufacture chips. But, in 1999 manufacturers began using lasers which functioned with ultraviolet light. In the ultraviolet light spectrum the beam's diameter was shrunk down to about a fifth of the 1 micron limit, and chips were produced that broke this "limit".

With the 2010 deadline looming, many companies, universities, scientists, and researchers are exploring other ways to form smaller, faster, more powerful computer chips. They have turned their efforts to ground breaking techniques — higher dimensions, nanotechnology, and molecular biology.

What do these three prospects have to do with computer chips? The elements imprinted on a silicon chip lie on the plane of the chip. In 1997 Ed Fredlin considered the possibility of stacking the circuits on top of one another on the silicon wafer. In other words, make the chip 3-dimensional. This would enhance the capacity of the chip many times over. Matrix Semiconductor in Santa Clara CA, Rebsselaer Polytechnic, SUNYAlbany, and teams of researchers at MIT and IBM are experimenting with these highrise chips.

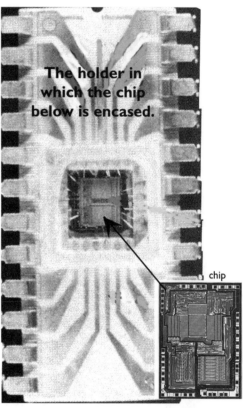

The holder in which the chip below is encased.

chip

For now the Matrix chips are primarily being designed to increase memory. These chips can be used in such devises as digital cameras, cell phones, and appliances. In addition, engineering and manufacturing techniques are also exploring these highrise chips for increased computing power rather than just increased memory.

In August of 2001 IBM announced what it considered a viable heir to the silicon chip — a chip composed of carbon nanotubes. Computer processing power is enhanced when transistors and logic circuits are shrunk, thereby making the electrons' paths shorter, faster and more energy efficient. Companies delving into the feasibility of using carbon nanotubes are hoping to develop ultra small chips to replace the silicon chips. "Intel can squeeze some 42 million of these transistors onto its Pentium 4 chip".[1] If nanotubes pan out as viable transistors, billions of these nanocomponents will be able to fit on a chip. Imagine how computer power and memory will escalate. But implementing new technology involves enormous start up costs. As the traditional silicon process becomes more difficult, perhaps the cost will outweigh the start up manufacturing expenses of nanotube transistors. Nanotubes have a few things in their favor — they may be able to be integrated into the existing computer framework, and they emit electrons at very low voltage. These nano-wonders are being considered for computer chip designs, but also for use in flat-panel displays with each pixel having its own personal emitter. Perhaps these will eventually replace the liquid crystal or plasma displays. Researchers are not only working with carbon nanotubes, but they have began delving into forming nanotubes out of other elements and exploring other uses. Researchers at Lucent Technologies' Bell Labs in New

Jersey surprised the computer-design world in November of 2001 with the announcement of their process of forming single-molecule transistors on circuit using a chemical solution in a bottoms-up formation process.

Molecular biology is a form of nanotechnology which utilizes nature's molecular sized machines. In particular, the properties of DNA and peptides are being experimented with and developed. Scientists[2] are seeking to harness the properties of how DNA nucleotide bases bond in hopes of forming viable rectangular DNA type circuit boards. Others[3] are working with peptides that bind to crystals in hopes of directing molecular transistor crystal growth of silicon. At Hewlett Packard and the University of California at Los Angeles researchers were successful in working with rotaxane molecules, making them form into a single molecular layer and possibly operate as transistors.

With computer scientists, molecular biologists, physicists, chemists, nanotechnologist, and mathematicians working in various directions, computer's power and memory should delay the 2010 predicted time barrier. Will they make the deadline?

---

[1]Nanotube Computer by David Rotman, Technology Review, March 2002)

[2] Ned Seeman and colleagues of NY University.

[3]Material chemist Angela Belcher of the University of Texas at Austin and physicist Evelyn Hu of the University of California at Santa Barbara.

the Internet & www

where, why & when

Today the Internet has become a household term. It's hard to believe that in the early 1960s it was merely an idea that would have such a phenomenal impact on how we communicate, do business, do research, be entertained, and even socialize. Who could have imagined the Internet would create a global seamless borderless world of its own in which you could access or send information to any part of the world or in some cases into outer space? The Internet's story is based on a sequence of inventions and ideas spanning centuries. Many crucial steps such as the development of binary numbers, the *memex*, the telephone, the computer, etc. were involved. Some people suggest that the USSR's launch of Sputnik in 1957 acted as an impetus for the United States Department of Defense to form the **A**dvanced **R**esearch **P**roject **A**gency (ARPA) which subsequently funded research and work for ARPANET which eventually became the Internet. The Internet would not exist without computers and telecommunication

The *memex* machine was a theoretical computer envisioned by Vannevar Bush (1890-1974) in the 1930s. He wrote an article, *As We May Think*, for the *Atlantic Monthly*, July 1945. In his essay he describes a futuristic device which surprisingly foreshadows the modern computer and the way its information is processed along the Internet today with hypertext linking. Bush writes *"The human mind ... operates by association. With one item in its grasp, it snaps instantly to the next that is*

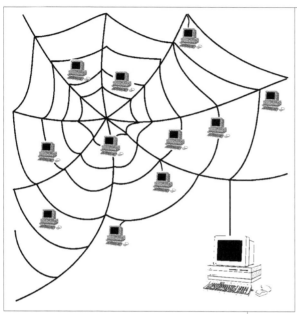

suggested by the association of thoughts, in accordance with some intricate web of trails carried by the cells of the brain. It has other characteristics, of course; trails that are not frequently followed are prone to fade, items are not fully permanent, memory is transitory. Yet the speed of action, the intricacy of trails, the detail of mental pictures, is awe-inspiring beyond all else in nature.

devices. Ironically, computers were initially invented to handle drudge work — number crunching at incredible speeds, data storage and retrieval, complex computational problem solving. Computers had not been conceptualized as communication devices. As with many inventions, new uses evolved, and so with the advent of the digital computer a new unexpected mode of communication also evolved. Computers and telephone networks were initially hooked up to allow researchers and scientists to tap into other computers and their number crunching powers. Since computers in the 1960s were still expensive and rela-

"...One cannot hope thus to equal the speed and flexibility with which the mind follows an associative trail, but it should be possible to beat the mind decisively in regard to the permanence and clarity of the items resurrected from storage.

"Consider a future device for individual use, which is a sort of mechanized private file and library. It needs a name, and, to coin one at random, memex will do. A memex is a device in which an individual stores all his books, *records, and communications, and which is mechanized so that it may be consulted with exceeding speed and flexibility. It is an enlarged intimate supplement to his memory. It consists of a desk...On the top are slanting translucent screens, on which material can be projected for convenient reading. There is a keyboard, and sets of buttons and levers. Otherwise it looks like an ordinary desk...."*

He later describes in detail how information is transferred from hard copy, stored and accessed. He explains how indexes are formed, how they can be easily linked and a "trail" formed which leads from one to another instantaneously by clicking a button. He even explains how more than one document can be viewed at the same time. His thoughts and ideas had a major impact on such designers as Douglas Engelbart and Ted Nelson.

tively scarce, being able to share time on a computer was an invaluable commodity for a researcher. In 1969 the ARPANET project set-up the first dedicated computer network through which scientists could share data bases and software. With the advent of ARPANET users also began sending short messages to each other. Computer networking pioneer Bob Lucky explains that "ARPANET was originally built for computers to talk to computers...But the computer operators started to send little messages back and forth between themselves...soon, other people started using this new mechanism to send little pieces of electronic mail from person to person through the network."[1] ARPANET initially set-up two *node*s — one at University of California at Los Angeles and the other at Stanford Research Institute(SRI). This connection allowed data and software to flow between the two computers. Later two additional nodes were installed at University of California at Santa Barbara and the University of Utah. By 1971 ARPANET had joined 15 nodes consisting of 23 hosts (computers). In 1972 Ray Tomlinson designed software which allowed e-mails to be sent between computers in ARPANET. This innovation was an almost instant hit. In 1983 the Internet emerged when ARPANET was split up into a military sector(called the Defense Data Network) and a civilian sector (NSFNet, a network of scientific and academic computers funded

In 1995, NSFNet began a phase withdrawal to turn over the Internet to a commercial consortium of providers, which include PSINet, UUNET, ANS/AOL, Sprint, MCI and AGIS-Net99.

by the National Science foundation). The ARPARNET was phased out in 1990, and the Internet was on its own.

The *Internet* can be viewed as an enormous network of networks which organizes and hosts these computer networks all around the world. Any of these subnetworks can send and receive data and share software. The vast Internet network manages volumes of information which is shared among the its networks and computers using a system called hypertext. The *hypertext system* allows information to be cross-

A *node* is a computer connected to a network. A network is formed when two or more computers are connected together thereby being able to share resources. When two or more networks are linked an *internet* is formed.

What's the difference between *internet* and *Internet?* *Internet* is an enormous collection of interconnected networks that are connected using the TCP/IP protocols which evolved from ARPANET. The Internet is probably the largest existing global internet connecting networks around the world.

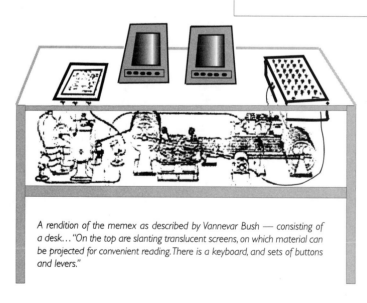

A rendition of the memex as described by Vannevar Bush — consisting of a desk… "On the top are slanting translucent screens, on which material can be projected for convenient reading. There is a keyboard, and sets of buttons and levers."

A *host* is any computer on a network which provides services to other computers on the network.

In 1971 Ray Tomlinson of Bolt Beranek and Newman, Inc. invented an e-mail program for sending messages between computers in a network. In 1972 he modified this e-mail program for ARPANET.

---

referenced interactively, allowing one to go effortlessly from one document to another. Today while researching information on-line, we take for granted the ability to simply click on a topic button, an underlined phrase, a highlighted word, or a picture and have new information immediately transported to our computers. The body of information and the hypertext system is what comprises the World Wide Web. The origin of the WWW dates back to 1989-90s when the first WWW software was written by Tim

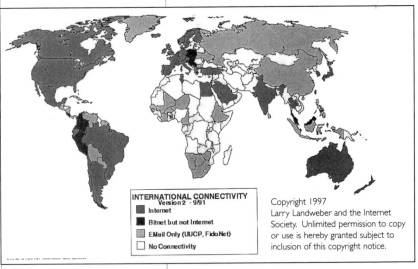

INTERNATIONAL CONNECTIVITY
Version 2 - 9/91
- Internet
- Bitnet but not Internet
- EMail Only (UUCP, FidoNet)
- No Connectivity

World Wide Web is technically defined as the collection of all resources and users on the Internet which use HTTP (Hypertext Transfer Protocol). This protocol moves hypertext files across the Internet. To function

Berners-Lee for sharing information on the Internet. While working at the CERN (Conseil Européen pour la Recherche Nucléaire), the European Laboratory for Particle Physics, and laboratories in Switzerland Berners-Lee and his colleague Robert Cailliau set up the first web server along with the needed protocols. The *Information Mesh, Mine of Information, Information Mine* were among the possible names they considered before settling for the World Wide Web. Berners-Lee describes the World Wide Web as "the universe of network-accessible information, an embodiment of human knowledge." The work of Berners-Lee and Cailliau encouraged

both the receiver and the sender must have this program on their computer.

In the 1960s Douglas Engelbart created the first *hypertext system.* The Internet protocols and the hypertext system are the foundation for the World-Wide Web.

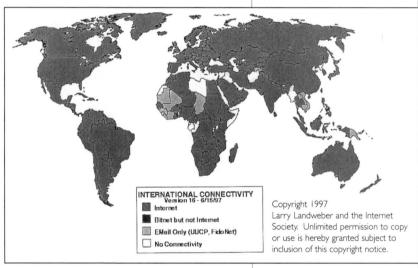

INTERNATIONAL CONNECTIVITY
Version 16 - 6/15/97
■ Internet
■ Bitnet but not Internet
▨ EMail Only (UUCP, FidoNet)
☐ No Connectivity

Copyright 1997
Larry Landweber and the Internet Society. Unlimited permission to copy or use is hereby granted subject to inclusion of this copyright notice.

Paul Kunz at Stanford Linear Accelerator to create the first US web site in 1991. By today's standards it looks like a bare-bones site — consisting of three lines of text, a link to get e-mail addresses and a link to search a scientific database. The data base consisted of an enormous number of archives of scientific papers,

It's the *protocol programs* that organize and manage the Internet's vast array of networks, computers, and functions.

TCP/IP — (Transmission Control Protocol/Internet Protocol) is the basic communication language of the Internet. For a computer to have access to the Internet, the TC/IP program must be installed on the computer. TCP handles the receiving, sending and reassembling of small packets of information. IP (Internet Protocol) handles the address of the small data packet, and insures it gets to the correct address. Every computer on the Internet has an IP number identifying where the data comes from and where it goes.

When a computer sends an e-mail or other form of information via the Internet, it is divided into small packets of data each identified with the address it originated from and with the address it is being sent to. These packets of information are transferred to gateways (computers) along the Internet. Each gateway receiving a packet sends it on its way to an adjacent gateway until it reaches a gateway which recognizes the address as a computer in its domain. The small packets are not kept in any order when they begin their voyage along the Internet, nor do they follow the same route of gateways. When they reach their destination, the TCl arranges the packets of data into the proper sequence before delivering the mail.

the perfect bait to attract researchers. The number of sites mushroomed so that by March 2002 the number had reached nearly 39 million. Similarly the number of hosts in ARPANET multiplied at a phenomenal rate so that from 1969 those two hosts at UCLA & SRI increased to nearly 148 million in January of 2002.

The Internet's structure is a type of open architecture allowing individual networks leeway in their designs. These networks conform to *Internet protocols* which enable all types of computers to both communicate and share services across all seven continents. The flexibility of the Internet to expand and touch all parts of the world and its diverse cultures makes it a global treasure and an incredibly powerful tool. The beginning years of ARPANET/Internet and the World Wide Web were fueled by individuals whose primary interest and focus were scientific and academic. Today in addition to these areas, the Internet is also being fueled by commercial and economic interests. How will the millions of laypeople and countless businesses tapping into WWW affect the Internet's evolution in the *next 20 years*? Will the same innovative drive to experiment and explore new methods and innovations be as creative and powerful? How will the impact of miniaturized Internet peripherals and new modes of Internet communications impact our daily lives? Will the Internet's influence draw more and more people into its sphere of influence and create an insatiable need to always be connected?

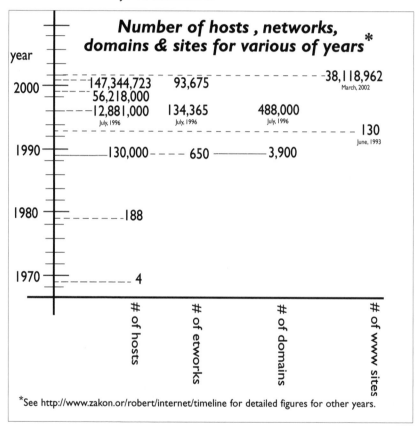

*See http://www.zakon.or/robert/internet/timeline for detailed figures for other years.

1p. 172 , *The Dream Machines* by Jon Palfreman & Doron Swade, BBC Books, London, 1991.

mathematics &
architecture
virtual architecture

Blobs and folds may not sound like exciting, sophisticated, or attractive adjectives to describe architectural forms. Yet these are the words being used to describe some of today's digitally generated buildings. These terms conjure up many of the bizarre shapes we see in one of the most contemporary and exciting 20th century structures — the Guggenheim museum in Bilbao, Spain by architect Frank Gehry. Architect Greg Lynn first used the term blob to describe computer generated architectural forms. A blob has no predetermined shape, but appears to bulge. A fold can be thought of as an inverted blob. But digital architecture is not confined to blobs and folds. In fact,

virtual architecture — architectural structures designed from their inception digitally via a computer.

Mathematically, *virtual architecture* can be viewed as a form of *topology* where the set of points comprising a structure are transformed into wild new shapes. *Topology* is often referred to as rubber sheet geometry because it deals with

Guggenheim Museum Bilbao—July 1997.
Photo by David Heald. © The Solomon R. Guggenheim Foundation NY.

it's not confined to any set of shapes. It is fluid architecture which can assume any shape, morphing according to the ideas inputted and manipulated on the computer by the architect.

The layperson is still unaccustomed to the "strange" shapes and architectural forms inherent in blobs and folds. Not rectangular or circular, they are initially jarring for some because they are not the Euclidean geometric forms that have defined architecture over the millennia. With today's technologies and materials, these innovative buildings need not remain figments of the architect's imagination. When Pier Luigi Nervi, engineering consultant for St. Mary's Cathedral in San Francisco, CA was asked what Michelangelo would have thought of St. Mary's Cathedral he replied that "He could not have thought of it.

characteristics of objects which remain unchanged when that object is stretched or scrunched. A cube and a sphere in topology are considered equivalent forms — either form can be morphed into the other by stretching or scrunching. Many of the forms one sees generated by virtual architecture are reminiscent of forms that have undergone topological transformations.

h2House: Multi-Functional Visitor & Demonstration Center for OMV Austrian Mineral Oil, Co., Schwechat, Austria. The building's interior is divided into two zones by a translucent fabric on which computer images are projected. Special scaffolding allows the exhibition area to be very flexible. State of the art computer simulation software is used to model, monitor and change the form and alignment of the solar vault to control shading devices, photovoltaic cells, and minimize energy use. Photo courtesy of Greg Lynn. Reprinted by permission of NODE@GLF.COM.

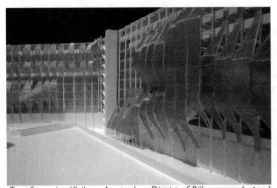

Transformation Kleiburg, Amsterdam District of Bijlmermeer. Instead of long horizontal corridors, vertical and diagonal ones are used to free up space for the units .The outer skin uses photovoltaic technology which covers ramps, escalators & walkways in a varying pattern. Photo courtesy of Greg Lynn. Reprinted by permission of NODE@GLF.COM.

Technically, Frank Gehry's structure in Bilbao is not virtual architecture because it was not conceptualized on a computer. He, instead, first designed a physical model, rather than a digital model of his conception. Then, using Catia, an advanced 3-dimensional modeler developed to map curved surfaces for the aerospace industry, Gehry was able to explore and alter shapes while remaining within the parameters of feasibility of constructible geometric forms and of materials utilized. Every point of the model's surface was mapped by using a digitizing process that traced the model's shape with a special arm-like digitizing tool. The information, when transferred to Catia, allowed the architect to explore shapes and maintain the geometric relationships to the constructibility of the model's shapes. While controlling a milling machine, Catia carved an exact model of the building's forms, maintaining dimensional control for the building's systems for the various materials used. Gehry points out that this new technology "provides a way for me to get closer to the craft. In the past, there were many layers between my rough sketch and the final building, and the feeling of the design could get lost before it reached the

This design comes from geometric theories not then proven." The same is true with revolutionary digital architecture. It could not have been created even a decade ago because the computers and software needed to carry out the engineering calculations and drawings did not exist. Computer programming innovations of the 1990s have made sleek, supple, natural, futuristic shapes and lines architecturally feasible. Until recently architects used computers only for the arduous and tedious work — the laborious architectural drawing, elevations, and engineering calculations. Now computers have become the tool of design of the virtual architect.

Unlike the architects of the past, the virtual architect no longer reaches first for a pencil and paper, but rather sits down in front of a computer monitor where the designs, shapes and forms from his/her imagination come to life. New software allows them to not only create 3-D renderings, but also easily generate perspective views and interior space. Even though virtual designs are very complex, the engineering calculations are accurate and structural calculations are carried out easily and quickly. In the past, an architect would labor for hours on a 3-D sketch or on changes in the rendering. Computer programming has simplified this work with changes done almost instantaneously as the

craftsman. It feels like I've been speaking a foreign language, and now, all of a sudden, the craftsman understands me. In this case, the computer is not dehumanizing; it's an interpreter." Yet his blobs, folds and curves have made a significant impact on the future of digitally designed architecture by being the first to introduce the general public to these new exciting shapes[1] — never before combined in a structure, and all of which would have not been possible if it were not for computer.

architect draws the structure. A digital design can be easily manipulated, twisted, flopped, stretched, and even perspectively altered. The computer gives the architect the freedom to experiment with the shape of the building. Virtual architecture allows the architect to explore many possible transformations or morphings of a structural idea while remaining within the parameters of the integrity and feasibility of the structure.

Besides being a whole new way to design structures, virtual architecture lends itself to a new way to view and experience designs before they have been constructed. In the past, miniature 3D models had to suffice for a client to view the architect's proposal. With virtual architecture the viewer not only sees the building on a computer monitor, but can virtually experience the building by walking through its different parts and levels. Digital buildings assume an almost life-like status.

What designs are popping up on computer screens? You can get a glimpse online of such structures as Greg Lynn's *Eyebeam Museum of Art & Technology* or

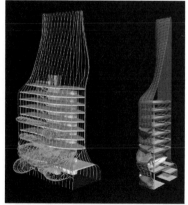

Eyebeam Atelier Museum of Art & Technology, is covered with an electronic skin making it into a new type of art /media/performance space. Photo courtesy of Greg Lynn. Reprinted by permission of NODE@GLF.COM.

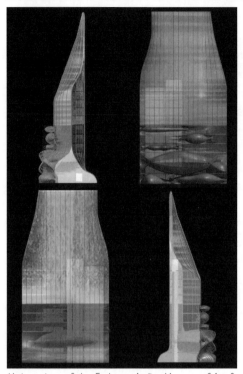

Various views of the Eyebeam Atelier Museum of Art & Technology. Photo courtesy of Greg Lynn. Reprinted by permission of NODE@GLF.COM.

his *Embryologic house.* Lynn does not like to destroy the continuity of a building's lines with windows and doors, so he uses movable panels. Look for the *Moebius House* by architects Ben van Berkel and Caroline Bos. In Paris experience first hand the new digitally designed restaurant in the Pompidou Center. Berkel and Bos have also been commissioned to do an addition to the Wadworth Atheneum in Hartford, Connecticut.

Architecture is a field whose evolution is defined and expanded according to the state of science, materials, and other technologies. Is it only its uniqueness that draws people or have its far out shapes also captured the public's imagination[2]. Will these shapes eventually become the norm as we become more accustomed and comfortable with them? It seems fitting that as the 21st century becomes more and more involved with computer technology so will its architecture.

---

[1]Seattle, Washington is the home of Frank Gehry 's *Experience Music Project (EMP)* which he designed for Microsoft co-founder Paul Allen. Its blobs and folds rival those of the Guggenheim in Bilbao, and is definitely a must see to believe. Its shapes are reminiscent of guitar parts consisting of six distinct but interconnected multi-curved forms covered with iridescent metals and painted aluminum.

[2]The Carnegie Museum of Art in Pittsburgh PA, featured digital architecture in its exhibit *Folds, Blobs, and Boxes: Architecture in the Digital Era.* (2/3/2001 to 5/27/2001)

mathematics'
highest award
the Fields Medal

I n 1994 John Nash, the brilliant mathematician portrayed in *A Beautiful Mind,* won a Nobel Prize for his ground breaking work in game theory. The prize was not given in his area of mathematics since there is no Nobel Prize awarded for mathematics. When the Nobel Prize was established upon the death and bequest of Alfred Nobel in 1896, a fund of $9,200,00 was allocated for annual Nobel Prizes in the areas of Peace, Literature, Physics, Chemistry, and Philosophy.[1] Alfred Nobel purposely omitted mathematics, and there has been much speculation about his omission[2]. Consequently, a mathematician can only receive a Nobel Prize by indirectly being recognized in one of the areas designated by the Nobel Prize. In the case of John Nash, it was given for his work in the area of economics.

In mathematics, the *Fields Medal* is equivalent to the Nobel Prize in prestige and recognition. Its formal name is the *International Medal for*

In 1981 the Executive Committee of the International Mathematics Union (IMU) established an award especially for mathematical aspects of information science (i.e.outstanding work in the fields of theoretical computer science). The following year the University of Helsinki offered to finance the award, and thus it came to be named the *Rolf Nevalinna Prize*. The cash award is donated by the University of Helsinki in memory of Finnish mathematician Rolf Nevalinna(1895-1980), who was Rector at the University of Helsinki and served as president of the IMU from 1959 to 1962. The IMU appoints a committee of three mathematicians to select the recipient of the Nevalinna Prize. The *Fields Medal* recipients are selected by a committee of eight mathematicians appointed by the IMU.

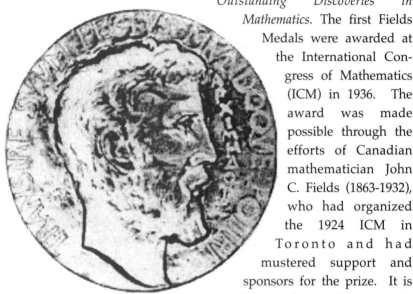

Fields Medal, front and back views

Outstanding Discoveries in Mathematics. The first Fields Medals were awarded at the International Congress of Mathematics (ICM) in 1936. The award was made possible through the efforts of Canadian mathematician John C. Fields (1863-1932), who had organized the 1924 ICM in Toronto and had mustered support and sponsors for the prize. It is bestowed every four years in connection with the ICM worldwide conferences. As John Fields mentions in his notes outlining the award, *"The hands of the IC should be left as free as possible. It would be understood, however, that in making the awards while it was in recognition of work already done it was at the same time intended to be an encouragement for further achievement on the part of the recipients and a stimulus to renewed efforts on the parts of others."*[3]

The medals are given to mathematicians who are no older than 40 in recognition of existing work and promising future work. Until 1966, two medals were awarded at each ICM conference, but due to the expansion of mathematical research into so many areas, each congress since 1966 has awarded four metals. The chart indicates the recipients since 1936. At the opening ceremonies of the 1998 ICM conference, Andrew Wiles was honored with a special tribute and awarded the first ever IMU (International Mathematical Union) *Silver Plaque* for his proof of Fermat's Last Theorem. Since Wiles was just over 40 when he completed his final version of his proof of Fermat's Last Theorem, he could not be awarded the Fields Medal, but the IMU created the *Silver Plaque* to recognize and honor Wiles for his outstanding achievement[4].

---

[1]In recent years, the Swedish government established the Nobel Prize for Economics, which is subsidized by the government and not by Nobel's fund.

[2]These speculations include — he did not like mathematics and felt it was an impractical science with limited applications and benefits for humanity — he held a grudge against mathematicians because a (false)rumor claimed his mistress had an affair with a mathematician — he had a falling out with mathematician Gosta Mittag-Leffler and wanted to insure there was no chance he would get the award.

[3]From the original letter by John Fields creating the endowment — see http://www.cs.unb.ca/~alopez-o/math-faq/node47.html.

[4] In addition, Wiles has received the MacArthur Fellowship, the Wolf Prize, and Clay Research.

Fields Medal Recipients

recipient	country	topics include	age
1936			
Lars Ahlfors	Finland	Riemann surfaces	29
Jesse Douglas	USA	Plateau problem	39
1950			
Laurent Schwartz	France	theory of distributions	35
Atle Selberg	Norway	sieve methods/Riemann zeta functions	33
1954			
Kunihiko Kodaira	Japan	harmonic integrals /Hodge manifolds	39
Jean-Pierre Serre	France	momotopy sphere group/spectral sequences	33
1958			
Klaus Roth	Germany	solved Thue-Siegel problem	32
Rene Thom	France	cobordism in algebraic topology	35
1962			
Lars Hormander	Sweden	partial differential equations	31
John Milnor	USA	7-dimensional sphere/differential topology	31
1966			
Michael Atiyah	UK	topology/ K-Theory/Index Theorem/Fixed Pt.Theorems	37
Paul Cohen	USA	proved independence of the Continuum hypothesis	32
Alexander Grothendieck	Germany	extension of theories, esp. on number fields	38
Stephen Smale	USA	research in topology	36
1970			
Alan Baker	UK	Gelfond-Shneider theorem/transendental #s	31
Heisuke Hironaka	Japan	generalized work of Zariski	39
Serge Novikov	USSR	topology/Pantrjagin classes	32
John Thompson	USA	extented work on non-cyclic finite simple groups	37
1974			
Enrico Bonbieri	Italy	primes/univalent functions/Berstein's problem	33
David Mumford	UK	problems of moduli/ theory of algebraic surfaces	37
1978			
Pierre Deligne	Belgium	Weil conjectures/algebraic geo.&number theory	33
Charles Fefferman	USA	multidimensional complex analysis	29
Gregori Margulis	USSR	Lie groups/combinatorics/diff.geo./dynamical sys.	32
Daniel Quillen	USA	high algebraic K-theory/ring& module theory	38
1982			
Alain Connes	France	theory of operator algebras	35
William Thurston	USA	revolutionized topology in 2 & 3 dimensions	35
Shing-Tung Yau	China	diff.eqs./Calabi conjecture in alg.geo.	33
1986			
Simon Donaldson	UK	topology of 4-manifolds	27
Gerd Faltings	Germany	proved of Mordell Conjecture	32
Michael Freedman	USA	proved 4-D Poincaré Conjecture	35
1990			
Vladimir Drinfeld	USSR	quantum groups & number theory	36
Vaughan Jones	N Zealand	discovered new polynomial. invariance for knots	38
Shigefumi Mori	Japan	algebraic manifolds/proved Hartsborne conjecture	39
Edward Witten	USA	discovred new symmetries in knot theory	38
1994			
Jean Bourgaim	Belgium	analysis/geo. of Banach spaces/convexity og higher dim.	40
Pierre-Louis Lions	France	contributions to theory of nonlinear part. diff. eqs.	38
Jean-Christophe Yoccoz	France	work in dynamical systems	37
Efin I. Zelmanov	Russia	solved Burnside problem/work in topology	39
1998			
Richard Borcherds	UK	proved Moonshine conjecture	39
Maxim Kontsevich	France	string & knot theories; proof Witten conjecture	34
William Gowers	UK	functional analysis; combination theory	35
Curtis McMullen	USA	work in complex dynamics/chaos theory	39
2002 (not announced as of this printing) to be held in China			

Leonardo da Vinci
Renaissance man

L eonardo da Vinci (1452-1519, a visionary whose work encompassed the arts and the sciences, is often referred to as the *Renaissance man.* If he were to submit a resumé today it would include — architect, scientist, painter, sculptor, musician, civil and military engineer, botanist, astronomer, cartographer, geologist, inventor, goldsmith with special interest and training in optics, hydrology, perspective, mechanics, and mathematics.

Leonardo created not only for his time, but often for the future. He designed canal systems, military fortifications, an armored tank, an early type of firing machine gun cannon. He spent hours studying the flight of birds, and how to incorporate his observations for human flight. In fact, his flying machine designs can be considered predecessors to today's hang gliders and his propelled machine to a helicopter. In his paintings and drawings he gave depth of field to his work with his mastery of perspective. He was driven to analyze subjects from the inside out. Witness his work in anatomy, from which he learned about bone

Leonardo, self-portrait.
From The Literary Works of Leonardo da Vinci by Sampson Low, Marston,
Searle & Rivington, London, 1883...

He envisioned humans as being able to someday fly, not only using their own bodies, but by using such mechanisms as his heliscrew, a prototype of today helicopter

.

Leonardo da Vinci emphasized the importance of understanding and studying mathematics as a framework to his various works and projects. In the late 15th century da Vinci drew a stallion which was to be a bronze sculpture towering 24 feet high. In the design of the horse he incorporated the golden rectangle's dimension to create dynamic symmetry. He then made a full sized clay model, which was destroyed by French troops invading Milan. Unfortunately, the sculpture was never cast. 500 years later, a foundation endowed by Charles Dent, had the sculpture recreated and two cast in bronze. One for a gift to Milan, which was unveiled on September 10, 1999, and the other for display in Beacon, Iowa.

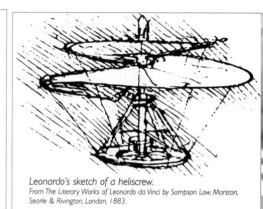

Leonardo's sketch of a heliscrew.
From The Literary Works of Leonardo da Vinci by Sampson Low, Marston, Searle & Rivington, London, 1883.

structure, muscles, ligaments, tendons, veins and arteries by personally undertaking numerous human dissections. Perhaps it was the enthusiasm and interest he put into what he undertook, or perhaps it was the scientific and mathematical analyses with which he approached his subject, which gave his work such a special quality. Regardless, his drawings and creations capture more than detail. They exude a spirit of their own.

Although many of his notebooks and books have been lost, of the voluminous notes he wrote over 7000 pages remain today — spanning his varied interests, including sketches, studies, findings, inventions, philosophy. In nature he worked at capturing the movement of water and waves, the

One of Leonardo's numerous horse studies.
From The Literary Works of Leonardo da Vinci by Sampson Low, Marston, Searle & Rivington, London, 1883..

movement of horses[1], the flight of birds, the form and structure of plants. In architecture, his designs included bridges, churches, cathedrals, canals. His inventions and their sketches illustrated his expertise in mechanics and physics. His work in optics and perspective led him to invent the perspectograph. He was not satisfied with merely producing works of art, but had an inherent need to understand the entire process, leading him to uncover and understand the mathematics behind things. He described his study of a bird as: "A bird is an instrument working according to mathematical laws, this instrument is within the grasp of man to reproduce with all its movements..." (from Codex Atlanticus)

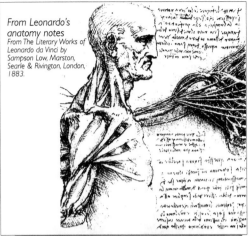

From Leonardo's anatomy notes
From The Literary Works of Leonardo da Vinci by Sampson Low, Marston, Searle & Rivington, London, 1883.

Perspectograph as illustrated by Leonardo in Codex Atlanticus

To Leonardo mathematics was part of all sciences, and consequently its knowledge and understanding were essential. He felt "No human inquiry can be called true science without going through mathematical tests..." His manuscripts after 1496 emphasize how he immersed

One of Leonardo's sketches illustrating how he used concepts of projective geometry to arrive at different views of this face. From The Notebooks of Leonardo da Vinci, Dover Publications, Inc. NY, 1970.

One of Leonardo's sketch for Adoration of the Magi illustrates his detailed use of perspective and vanishing points. From one of his many his notebooks.Uffizi Gallery, Florence, Italy.

himself in geometry and Euclidean problems. Pages upon pages of his notebooks were filled with mathematical diagrams, problems, solutions. He studied Euclidean geometry, worked with symmetries,

As an artist he may be best known for his Mona Lisa (or La Gioconda, the wife of the merchant Gioconda) and his mural The Last Supper. From The Notebooks of Leonardo da Vinci, Dover Publications, Inc. NY, 1970.

and pondered over geometric problems, explored the golden ratio, studied curves with single and double curvature, stellar polygons, constructions of regular polygons, and applied mathematics to optics and physics. Leonardo utilized mathematics and mathematical forms all aspects of his art and embraced a sense of symmetry and proportion. He wrote, "Let proportions be found not only in numbers and measures, but also in sound, weights, times,

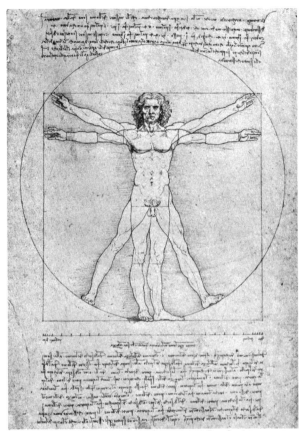

Leonardo's study the the proportion of the male body. From De divina propotione.

positions, and whatever force there is." (from Manuscript K). His illustrations in the book, *De divina proportione* written by Franciscan monk and mathematician Luca Pacioli, show both his mastery of the Platonic and Archimedean solids and elaboration on the proportions of the body as first outlined by Vitruvius, a Roman architect of 1 B.C.). In *De divina proportione*, his famous man with outstretched arms appears, suggesting how the human body is proportioned. The figure's design is based on a square and circle — the square surrounding the man illustrates

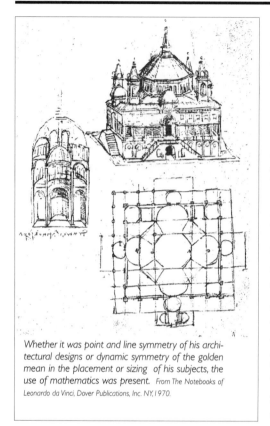

Whether it was point and line symmetry of his architectural designs or dynamic symmetry of the golden mean in the placement or sizing of his subjects, the use of mathematics was present. From The Notebooks of Leonardo da Vinci, Dover Publications, Inc. NY, 1970.

that his height is equal to his outstretched arms, and the four arms and legs show how the figure is circumscribed by a circle. When the arms are raised so that the middle fingers of the hands reach the head's height and the legs are spread apart so that the figure's height is decreased by 1/14, then the outstretched legs and risen arms touch four points of the surrounding circle. In addition, an equilateral triangle is formed by the two parted legs. Regardless of which of Leonardo's work we view, we find mathematical ideas — whether it was his sketched horse, an architectural design, or an invention — mathematics was a key factor.

---

[1] In his sculptures, he captured not only the detail, the gait of a horse, but also its spirit.

What's it all about?
bifurcation, periodic doubling and the Feigenbaum constant

O ver 5,000 years ago, an amazing mathematical discovery was made by ancient "mathematicians" who were intrigued by circles. Drawing circles of all sizes and measuring the distance around them, they discovered that the ratio of a circle's circumference to its diameter was always the same regardless of the size of the circle. That constant came to be called pi, and its symbol, π, was first used in 1706. Similarly the constant $e \approx 2.71$ was discovered, and was given its name e by Leonhard Euler in the 1700s. Today e is used in describing the half-lives of radioactive materials, is hidden in the fibers of an orb spider's web, and is even used to compute perpetual interest rates on bank accounts. When these two constants were discovered who would have imagined that their impact would be so significant?

Fractals, chaos and complexity , like all mathematics, are interrelated. In essence, a fractal is simply formed by starting with an object and rule. The rule is applied to the object, and a new similar object emerges. Then the rule is reapplied to the new object, and again a new similar object is formed. This reapplying of the

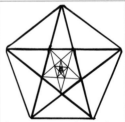

A pentagon can be used to generate a fractal by drawing in it diagonals and forming a pentagram. Within each pentagram is another pentagon in which its diagonals can be drawn to produce another pentagram. Each generation is similar in shape to the previous ones, and the process can go on infinitely.

rule to each new object to give again a new one produces the evolving fractal. The object can be almost anything, such as a regular pentagon *with the rule:* draw in its diagonals. Or the object may be a number (such as the number 0.5) and an equation (called an iterative equation). For example, consider the iterative equation:

$$x_{new} = 3 \, x_{initial}(1-x_{initial}).$$

The constant 0.5 is replaced for $x_{initial}$ and we get 0.75 for x_{new}. Then 0.75 is replaced in the equation for $x_{initial}$ and 0.5625 comes out, and the process is continued forever.

bifurcation is splitting or forking of a phenomenon.

periodic doubling — when the occurrence of a phenomenon begins occurring twice as often, then four times as often, then 8 times as often, then 16 times, etc.

What do generating fractals have to do with bifurcations? In the mid-1970s physicist Mitchell Feigenbaum, at the Los Alamos Labs in New Mexico, first looked at a very basic iterative equation, namely $x_{new} = k \, x_{initial}(1-x_{initial})$. In the example above 3 was selected for k, but k can be given any value. Once a beginning value for $x_{initial}$ is chosen, the equations generates infinitely many numbers. Using only a programmable hand calculator, Feigenbaum generated the values when k was between 1 and 4. He particularly studied what occurred as starting values for $x_{initial}$ were selected closer and closer to 0. He began comparing successive ratios from the results generated by the iterative equation. To his amazement two things were happening that were somehow related. A single outputted result stopped occurring after a particular point, and instead, a split occurred and 2 results were generated. Then these 2 results eventually gave 4 results. Then these 4 gave 8, and on and on. In mathematical terms, he was getting *bifurcations* (splitting results) and *period doubling*. Then when he examined the ratios between the bifurcations of the period doublings, he found that they would always approach the same number regardless of the values he started with. This number has come to be called *Feigenbaum*

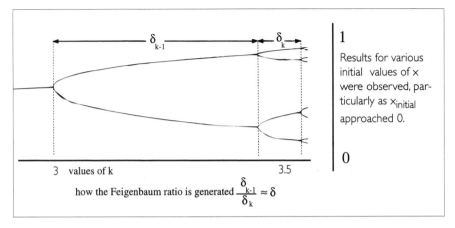

how the Feigenbaum ratio is generated $\dfrac{\delta_{k-1}}{\delta_k} \approx \delta$

constant, δ, 4.449211660910299067... To his amazement he also found that different iterative equation's ratios also yielded this same number. Even iterative equations that didn't seem to resemble one another, for example $x_{new} = k \sin(x_{initial})$, produced this ratio. *What did this mean?* It meant he found a *universal constant,* a number that was common among a class of chaotic systems, similarly to how π is a common number for the class of geometric objects known as circles. We know that π is used to calculate a circle's perimeter and circumference, but how is this Feigenbaum number used?

For complex systems which bifurcate and have period doubling the Feigenbaum number indicates when to expect the next bifurcation, and the next and the next; in other words when to expect a complex system to become chaotic. Applications of the Feigenbaum constant and others that may appear with other classes of complex systems can be applied phenomena in the universe. For example, the constant can be used in such areas as population growth, the spread of disease, the turbulence in weather and fluids, the statics in electronics, the flow of energies, the movements economics, biological and chemical systems, and many other aspects of the universe.

higher dimensions
a glimpse at how mathematics explores higher dimensions

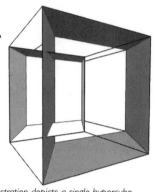

I t does not seem possible that mathematicians can explore objects and ideas which cannot be seen, but eyes of mathematics are not confined to what we see but rather to what we think and where this thinking leads. Ever wonder how mathematicians and physicists come up with such far out ideas about the essence of matter? For example, there are theories which consider multi-dimensional vibrating strings to be building blocks of the universe in the Theory of Everything (TOE). These are considered to be compacted in higher dimensions. Naturally, no one can see these strings or physically experience these higher dimensions. Now there is a revised, updated form of TOE called M-Theory. It reconciles the mathematical inconsistency of past superstring theories and actually is

The above illustration depicts a single hypercube rotating in space. The illustration below shows how the hypercube unfolds in the 3rd dimension.
Images courtesy Davide P. Cervone, Dept. of Mathematics, Union College, ©Copyright 2002.

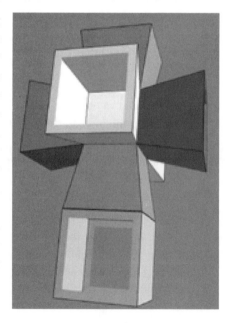

believed to encompass them in its domain. M-theory works with 11-dimensional space in which its strings are actually membranes crammed into these higher dimensions. M-Theory, which has been evolving since 1998, seems to reconcile the various forms of the superstring theories into itself.

But how does the mathematics behind such ideas work? An eye opening example makes use of the ancient Pythagorean theorem to explore higher dimensions. This gives insights and a glimpse into how mathematics is used to explore higher dimensions and its mathematical problems.

Consider a 4" by 4" square enclosing 4 congruent circles each with 1" radii.

Suppose a fifth circle is placed between the 4 circles, as shown in the diagram below.

Notice that this circle is smaller and how it is packed inside the square. What is this circle's diameter? Mentally draw in the diagonals of the square. Where they intersect is the center point of the square.

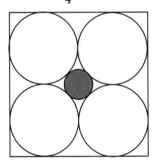

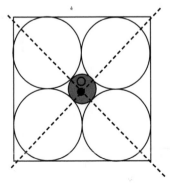

Through this point, O, set up the x-y axes with origin at this point. The small circle has its center at this origin. The centers of the other four circles are located at (1,1); (1,-1); (-1,1); (-1,-1) as shown.

Here the Pythagorean theorem enters the picture, and applying the theorem the radius of the small circle is determined as follows.

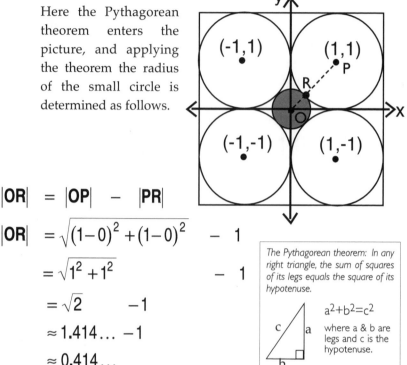

$$|OR| = |OP| - |PR|$$

$$|OR| = \sqrt{(1-0)^2 + (1-0)^2} - 1$$

$$= \sqrt{1^2 + 1^2} \qquad - 1$$

$$= \sqrt{2} \qquad -1$$

$$\approx 1.414\ldots -1$$

$$\approx 0.414\ldots$$

> The Pythagorean theorem: In any right triangle, the sum of squares of its legs equals the square of its hypotenuse.
>
> $a^2 + b^2 = c^2$
>
> where a & b are legs and c is the hypotenuse.

This value makes this circle's diameter about 0.82 inches.

Now carry the square into the next dimension, namely, the 3rd-dimension. Here it is a cube, and the 4 circles are replaced by 8 spheres. Again, use the Pythagorean theorem to determine the radius of a smaller 9th sphere packed in the center of the eight spheres. The centers of the 8 spheres are located in the

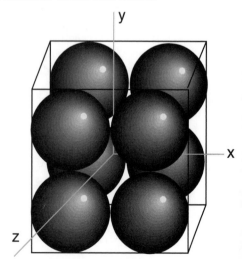

x,y,z space at (1,1,1); (1,1,-1); (1,-1,1); (1,-1,-1); (-1,1,1); (-1,-1,1); (-1,1,-1); (-1,-1,-1) while the 9th sphere's center is at the origin (0,0,0). One of the beauties of the Pythagorean theorem is that it applies to any dimensional space. So now, the 9th sphere's radius is determined in the same

$$\sqrt{(1-0)^2 + (1-0)^2 + (1-0)^2} - 1$$
$$= \sqrt{1^2 + 1^2 + 1^2} - 1$$
$$= \sqrt{3} - 1$$
$$\approx 1.732\ldots - 1$$
$$\approx 0.732\ldots$$

fashion as with the circles. It comes out to be approximately 0.732.

We see a pattern emerging for the radius of any dimension n. Here an nth-dimensional hypercube is packed with n nth-dimensional hyperspheres. And in each case, the radius of the small inserted hypersphere can be determined by

$$= \sqrt{1^2 + 1^2 + 1^2 + \ldots + 1^2} - 1$$

note: 1^2 is added n times

But, when exploring the radii of all these little inserted hyperspheres, something amazing happens. Applying our equation in the 10-dimension space, we get its radius to be about 2.162, which means its diameter is about 4.324. This means the little inserted 10-dimensional hypersphere pokes outside the 10-dimensional hypercube since the cube's dimensions are 4 by 4 by 4 by 4 by 4 by 4 by 4 by 4 by 4 by 4.

$$= \sqrt{1^2 + 1^2 + 1^2 + 1^2 + 1^2 + 1^2 + 1^2 + 1^2 + 1^2 + 1^2} - 1$$
$$= \sqrt{10} - 1$$
$$\approx 3.162\ldots - 1$$
$$\approx 2.162\ldots \quad = \text{the small hypersphere's radius, which makes its diameter} \approx 4.324.$$

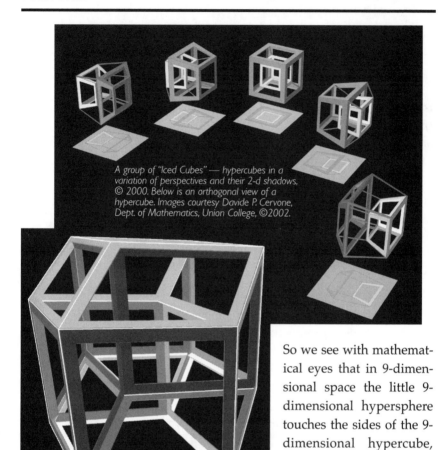

A group of "Iced Cubes" — hypercubes in a variation of perspectives and their 2-d shadows, © 2000. Below is an orthogonal view of a hypercube. Images courtesy Davide P. Cervone, Dept. of Mathematics, Union College, ©2002.

So we see with mathematical eyes that in 9-dimensional space the little 9-dimensional hypersphere touches the sides of the 9-dimensional hypercube, and beyond 9 dimensions it punctures its sides.

In 100-dimensional space, we get its radius to be √100 – 1 = 10 – 1 = 9. Thus, the hypersphere's diameter is 18, while the 100-dimensional hypercube that it was placed in has dimensions 4 by 4 by 4 by ... by 4.

Can you envision this? Mathematics can — and says it's so!

mathematics & population

What's the connection?

T he new millennium brings with it old problems and hopefully new solutions. Population and its impact on the earth is one such problem that dates back centuries.

Around 500 B.C. Han Fei-Tzu of the Chou Dynasty wrote *"In ancient times, people were few but wealthy and without strife. People at present think that five sons are not too many, and each son has five sons also and before the death of the grandfather there are already 25 descendants. Therefore people are more and wealth is less. They work hard and receive little.*

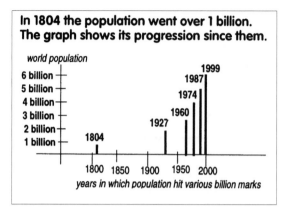

**In 1804 the population went over 1 billion.
The graph shows its progression since them.**

world population

years in which population hit various billion marks

The life of a nation depends upon having enough food, not upon the number of people."

Circa 160-230 A.D. Tertullian wrote in *De Anima, "The strongest witness is the vast population of the earth to which we are a burden and she scarcely can provide for our needs; as our demands grow greater, our complaints against nature's inadequacy are heard by all. The scourges of pestilence, famine, wars, and earthquakes are heard by all, regarded as a blessing to overcrowded nations, since they serve to prune away the luxuriant growth of the human race."*

In 1755 Benjamin Franklin wrote *"There is, in short, no bound to the prolific nature of plants or animals, but what is made by their crowding and interfering with each other's means of subsistence."*

In 1798, Thomas Robert Malthus wrote his *Essay on the principle of Population* in which he stated *"Population, when unchecked , increases in a geometrical ratio. Subsistence increases only in an arithmetical ratio. A slight acquaintance with numbers will show the immensity of the first power in comparison of the second."*

And in the 20th century, Paul Ehrlich observed *"Everyone understands that the population explosion is going to come to an end. What they don't know is whether it's going to come to an end primarily because we humanely limit births or because we let nature have her way and the death rate goes way up."*

According to United Nations revisions of world population estimates and projections, figures show—

• The world population reached the 6 billion mark in 1999, which is an average of 214,986 people per day growth.

• The global average fertility level in 1999 stood at 2.7 births per woman compared to 5 births per woman in the early 1950s.

•HIV/AIDS epidemic demonstrates devastating mortality tolls[1].

• Longevity is expected to increase. In 1998, one person in every hundred was 80 or over, which totals about 66 million. By 2050 these figures are expected to be 6 times higher, reaching a total of 370 million worldwide.

As of July 2002 the African nations Botswana, Zimbabwe, Swaziland, Lesotho, Nambia, Zambia and South Africa had HIV rates in excess of 20% for people 15 to 49 years old.

$$\sum_{n=1}^{\infty} \left(\sqrt{\frac{land}{resources}} + \left(\frac{1}{human\ consumption} \right)^{1-population} \right)^{n}$$

Naturally, this is a bogus equation, but mathematics is developing tools to deal with population problems involving everything from insect populations, to bacteria, to viruses, to elephants, to rain forests, to gray whales and to people. Just considering any one of these groups in a fixed environmental space is difficult. Now imagine the complexity of dealing with populations of all things on the entire earth all interacting in their environments simultaneously, and then stir in their interdependence.

Mathematicians know that population growth and our ability to predict its trend is influenced by many factors. Among these are epidemics, disease, natural disasters, war, droughts, famine, social and religious mores. Add to these fertility and consumerism trends, and we quickly realize the mathematics needed to predict population growth cannot be a linear equation, not even a system of linear equations with specific restrictions and factors; nor Malthus' 1798 geometric and arithmetic progression descriptions.

> In a *linear system* the same conditions or values always produce the same results. In other words, it is predictable. On the other hand, a *non-linear system* is dynamic. The same conditions may produce entirely different results. The elements of complexity and chaos are inherent in such a system.

We know Malthus' mathematical view of human population growth as tied to a geometric progression was flawed. Among his oversights, his model did not take into consideration population sprawl; nor did its arithmetic subsistence level consider the possible advances in technology and agriculture. His model could be used to describe a severely restricted hypothetical population. For example, suppose a certain plant dies off in the summer and triples[2] each winter. If S_1 is the size of its population in the first year, then the next year's population is given by $S_2=3S_1$, and subsequent years are given by $S_{t+1}=3^{t-1}S_t$, (where t stands for time increments) t= {1,2,3,4...}, barring the introduction of pests such as snails which might devour many plants. Its geometric progression would be — S_1, $3S_1$, $9S_1$, $27S_1$, $81S_1$, $243S_1$, $729S_1$,...,$3^{t-1}S_1$,.... . In 10 years the plant population would be 3^9S_1. In 1845 Pierre François Verhulst, a Belgian mathematician, modified this type of geometric progression by introducing into it a new factor, namely $(1-S_t)$. With this addition, the formula becomes $S_{t+1}=3^{t-1}S_t(1-S_t)$. Suppose S were redefined not to represent the actual size of the population, but a fraction of the population (lying between 0 and1). This redefinition with the inclusion of the Verhulst factor, besides describing a population in a given year to be proportional to that of the previous year, creates a *feedback* factor in the equation. The feedback factor introduces concepts of iteration and nonlinearity into the formula, which brings into the picture the possibility of chaos. And it's chaos that inhibits a model's predictability.

Today, such mathematical work is known as *population dynamics.* A population dynamics model is a formula for a given biological species that predicts its population development growth over time. Even such simple formulas, as those mentioned, produce fractal pictures of systems which exhibit swings between order and chaos. The *strange attractor* and the *butterfly effect* find their way into these population problems where minute, even imperceptible changes in initial conditions can affect the outcomes quite rapidly. Since these population problems produce elements of complexity oscillating between episodes of order and chaos, imagine the results for complex systems such as that of Earth and the shortcomings of long range predictions. Here the types of problems are varied and diverse. One situation might deal with a confined population — another with the same population exhibiting fertility problems — or perhaps with a population that is not confined but has space to sprawl, while impacting its environment with a high degree of consumerism and little technological advances to compensate for its consumption level. Even if a population has 0 growth rate its consumption rate can deplete its resources thereby mimicking a rapid escalating population growth rate. Or consider modeling a rapidly growing population whose consumption is in perfect balance with its resources and with space to sprawl. Now factor in fertility, epidemics, wars, increased longevity, famine, natural phenomena/disasters, and the interdependence of these factors...these situations result in a gargantuan number of varied scenarios and diverse combinations. In all these cases there are the fixed factors of the Earth's finite size and resources. Many argue that expansion into outer space and future technologies will ultimately come to the rescue. But can technology always bail us out? Is there no limit to its ability to expand resources to enhance food and water supplies? Perhaps mathematics can give us an edge in preparing for these different scenarios, mapping out as many scenarios as our minds can conjure up.

What mathematics points out graphically is that the Earth's population and resources are in constant flux — a state of continual change. This flux has inherent in it elements of chaos. Mathematics can produce a host of varied models of hypothetical population scenarios of the earth under a myriad of changes. Which of these will actually materialize is, by today's mathematics, unpredictable. Perhaps advances in complexity theory, probability, statistical analyses will make breakthroughs in mathematical modeling and its ability to predict accurately further and further into the future.

---

[1]Although Botswana's population may be 23% smaller by 2025 than it would have been in the absence of AIDS, its population is expected to double between 1995 and 2020 because of its high fertility rate.

[2]The growth rate of 3 was just used for this example. Replace this with g in general form.

the nanos are coming
What's happening in nanotechnology?

n the 17th century Galileo wrote *"...if one wants to maintain in a great giant the same proportion of limb as that found in an ordinary man he must either find harder and stronger material for making the bones, or he must admit a diminution of strength in comparison with men of medium stature; for if his height be increased inordinately he will fall and be crushed under his own weight. Whereas, if the size of a body be diminished, the strength of that body is not*

> Engineers refer to Galileo's observation as the *cube-square law* — as the weight of an object increases the stress on its structure increases with the cube of its linear size, but their strength increases only by the square of its linear size.

diminished in proportion: indeed the smaller the body the greater its relative strength."[1]   Today over 400 years since Galileo made his observation, scientists have focused their attention to the world of the ultra small seeking to harness power and strength in new minute machines and devices.

1,000,000,000

Are the nanos actually coming? No! They are already here. The prefix nano- is being attached to everything from a wire to a barcode. Nanotechnology and nanofabrication deals with the world of the invisibly small. A *nanosomething,* whether it's a nanometer or a nanosecond, is defined as a billionth its size. So we are talking about machines and devices made up of a few atoms and manipulated using relatively new microscopic technology — such as micro twisters, scanning tunneling microscopes (STM), scanning probe microscopes (SPM), atomic force, microscopes (AFM) and magnetic resonance force microscopes (MRFM) microscopes — are making it feasible to move atoms around. Using knowledge of molecular structures coupled with these advanced technological tools, researchers push, probe and relocate atoms and molecules.

Where does mathematics appear in nanotechnology? It appears just about everywhere. It provides the numbers to measure and describe nanostuff, produces virtual animation on how such mechanisms would function, does calculations and provides equations describing behavior of nanomachines, looks to knot theory to explore the shapes which nano stuff can assume, and uses complexity and chaos theories to explain possible actions and problems of nanounits. Mathematics has a very big role in the very small world of nanotechnology. Without mathematics it would be futile to even consider delving into it.

Where did researchers come up with the idea of making things on such a minute scale, or believe such things would be useful? Researchers looked to things appearing in nature and the sciences of biology and chemistry, and observed how atomic sized objects functioned as tools and machines. They noticed how these elements worked, organized and controlled things around them. Our bodies function because of such minute organic devices. Consider protein molecules in the human cell which manipulate atoms in a substance and replicate themselves using codes and programs encrypted in DNA and the switches which genes control. Today proteins act as models which inspire researchers to seek out the possibilities of manufacturing such things as bearings, transistors, and computers — all designed at nanoscale. The chemistry of atomic structure explains which atoms combine with one another, how to join together different atoms, and how to explain their interactions. Perhaps the way living cells make copies of themselves may eventually help researchers make self-replicating nanomachines. Scientists are approaching the problem of manufacturing nanostuff in two ways — top-down approach or the bottom-up method. The top-down works at miniaturizing existing devices until they are nanoscale, while the bottom-up seeks to build nanounits from scratch using atoms as their building blocks. Physicist Richard Feynman first envisioned this approach with machines designed to make ever smaller machines over and over again until they were nanomachines. On the other hand much of the work in today's nanotechnology is following the route envisioned by Eric Drexler, who imagined working directly with atoms and molecules and building ultra small machines from scratch. This all may seem far fetched, but scientists have already made headway.

Less than a decade ago nanomachines and nanotools existed only as computer simulations. Today among the list of nanotools and nanodevices we find —

nanotubes — Tubes whose cylindrical wall is only a single atom thick and measures anywhere from four to six atoms in diameter. Researchers have fashioned them from various atoms, but by far the most popular have been the carbon nanotubes made from fullerenes. These ultra small tubes were first discovered in soot. Today scientists are using them

Fullerene is named after Buckminster Fuller because of its geodesic shape. The *buckminsterfullerene* has the shape of a soccer ball, consisting of 12 pentagons and 20 hexagons, with its chemical formula being C_{60}.

In 1985 the idea of nanotubes emerged with the discovery of *fullerenes* by Wolfgang Krätschmer at the Max Plank Institute for Nuclear Physics in Germany, where they found a way to mass produce fullerenes. In 1990 Sumio Ijima at NEC Research in Tsukuba, Japan discovered tube-like carbon structures — long carbon molecules nestled within one another which appeared as stacked graphite cylinders whose ends were capped with hemispheric fullerene. Then in 1993 nanotubes one atom thick were discovered simultaneously by both Ijima and the group of researches at NEC Research and a research team at the IBM Research in San Jose, CA. These tubes were only a few nanometers in diameter, and have come to be known as *carbon nanotubes* or *buckytubes*. From then on nanotubes mania caught on.

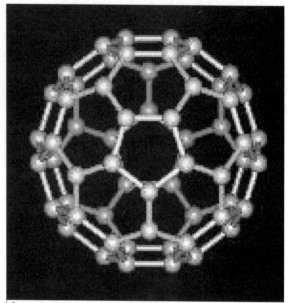

C_{60} molecule is an example of a fullerene, which is a caged molecule in the shape of a polyhedron and composed of carbon atoms. Courtesy of Professor Riichairo Saito, Japan.

in a variety of ways. They are working with these to make novel drug delivery systems by taking advantage of the space within the tube's formation. By designing carbon nanotubes of specific sizes and thereby allowing only specific molecules to enter, they coax the molecules of a particular drug into the tubes and direct the tube to the specific site of the body that requires treatment.

In the near future, TVs and computer monitors may be using nanotubes to form their screens. Samsung has nearly finished work on technology for flatscreen TVs and computer monitors based on carbon nanotubes. Each pixel would be connected to a carbon nanotube. These new displays will be less expensive than the current LCD(liquid crystal displays) or plasma display. In addition, they will be far more energy efficient than the cathode-ray tubes.

Then there are the researchers using nanotubes in innovative ways to create computer memory. A row and column of nanotubes are placed on parallel planes over one another to create transistors for a nonvolatile computer memory. Tom Rueckes, cofounder of Nantero company in Woburn, MA, is the inventor of this novel use of nanotubes. The design uses "an array of parallel nanotubes suspended just a few nanometers above a perpendicular array...each intersection represents a potential bit of memory...an applied electronic force stretched a tube on the top array close enough to the a lower tube, they physically bind and a current can flow between them; the switch is on and stays on even when the power is turned off."[2] This means a computer with this type of DRAM(dynamic random access memory — the memory that starts up the computer and runs its system and programs) will never need to be booted up. Nonvolatile memory is not lost when the computer is shut down. What's equally amazing is that a conventional sized centimeter chip

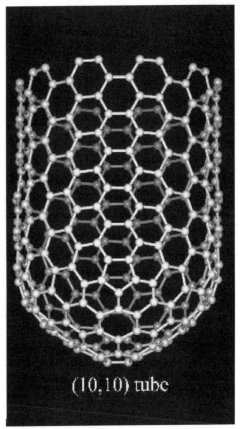

(10,10) tube

Part of a molecular model of a carbon nanotube.
Courtesy of the Smalley Research Group, Rice University, TX.

with this design may be able to hold a trillion bits. Carbon nanotubes are also being explored by IBM and other companies for use as transistors.

While physics graduate John Cumings at UC Berkeley was working on telescoping multiwalled nanotubes, a connection broke. The nanotubes layers snapped back into closed telescope formation. This led him to consider using these multiwalled nanotubes as springs which worked with nearly no friction, meaning its life span is enormous. Is this a new nanotool?

In addition, multiwalled nanotubes are being explored as frictionless springs. Nanotubes may eventually be used as chemical and biological detectors, targeted to combat drugs and terrorism.

How about lighting provided by nanotubes? Yes, nanotubes may by used in the future to create light bulbs using nanotubes as metal wire. When a voltage is applied nanotubes emit electrons which strike the phosphorescent coating of the glass cylinder. These lights can come on instantly, be easily dimmed, and do not give off toxic mercury vapor. The downside right now is how to design them to use less energy than a fluorescent light bulb. Jean-Marc Bonard of Ecolé Polytecnique de Fédérale de Lausanne in Switzerland and his team worked on this innovative way to use nanotubes.

dendrimers —These are molecules which grow like trees, and are based on fractal replication of polymers. The first molecular tree was fashioned from small organic molecules in 1978 by Fritz Vögtle and his colleagues at the University of Bonn in Germany. Today scientists are exploring ways to use the cavities inherent in the structure of dendrimers to transport drugs. The dendrimer cavities can accommodate medicinal molecules which are then transported by the dendrimer. "Model calculations have demonstrated that a dendrimer of six layers could take 10 to 20 molecules of the drug dopamine and deliver it to the kidneys. Unwanted side effects that the drug could trigger in the brain can be excluded, as the bulky transporter (unlike the naked dopamine molecule) would not be able to penetrate into the central nervous system."[3]

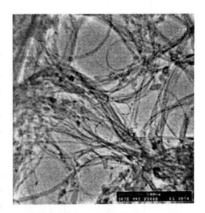

Nanowires have been developed and will be designed to work with electronic and optical devises. They are grown with stripes of different widths which indicate various semiconducting materials. Striped nanowires research is being done at Harvard University, University of California at Berkeley and Lund University in Sweden. A major focus is to use these nanowires to build memory mini chips, circuits, hybrid chips, and other devises.

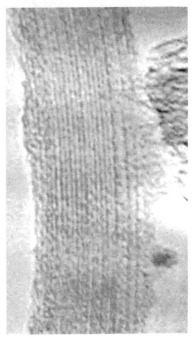

nanobarcodes — These have been made using nanowires with stripes of gold, silver and other metals. By varying the number, width and order of the stripes a host of different barcodes can be generated. Currently fluorescent markers used in biological tests can analyze a few molecules at once. The new nanobarcodes will

Both photographs are bundles of nanotubes, but in the lower photo the tubes are alined by intermolecular forces. Courtesy of the Smalley Research Group, Rice University, TX.

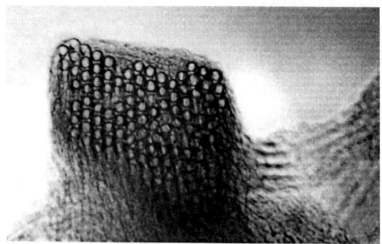

Nanotube-like ropes. Courtesy of the Smalley Research Group, Rice University, TX.

allow thousands of molecules to be tagged and identified. These nanobarcoded molecules will be used to identify the molecules of a disease. SurroMed of Mountain View, CA is exploring nanobarcodes to identify molecules in a diabetic's blood and the brain fluid of patients with Alzeimers. Nanobarcodes are even being placed on nanowires to identify the wire's semiconducting materials.

nanothermometers — These were invented by Yihua Gao and Yoshio Bando of the National Institute for Materials Science in Tsukuba, Japan. They will be used to measure temperatures in microscopic environments. The thermometer is composed of a nanotube filled with gallium which expands in relation to temperature changes. These thermometers are so small that an electronic microscope is needed to take the reading which can range between 50°–500°.

nanoears —These nanofibrils are fashioned after the minute hairs of the ear, called stereocilla. Stereocilla pick up sounds by sensing the vibrations created by the sounds causing them to vibrate just a few nanometers. Using nanotubes, researchers at Brown University in Providence,RI, are designing nanosized acoustic sensors which convert vibrations to electrical signals. These fibrils would be thousands of times smaller than stereocilla. They hope they will be used to improve hearing aids, detect sounds on interstellar travel, seek out cancer cells in the body, or as sound detectors on MEMS(microelectromechanical systems).

nanobelts — Nanobelts, ribbon-like nanostructures, are being made from semiconducting metal oxides. The belts synthesized by chemist William Buhro at Washington University in St. Louis are anywhere from 30 to 300 nanometers wide and about 10 to 15 nanometers thick which can be bent in half. There are various ways these belts could be used, such as in microscopic devices and as components of nanosensors. Further, they can be made from different oxides such as tin, zinc and lead, resulting in belts with similar ribbon structure.

nanobots — Fractals, by their nature, self-reproduce, and their reproductions are self-similar which can become smaller and

Thus far commercialization and globalization has eliminated about 75% of the Earth's natural genetic crop diversity. *Genetically modified crops* can adversely affect the remaining 25%. These *transgenic* crops were engineered using genes from other species. These genes were introduced into a crop's genome in hopes of making them hardier, and improving their survival by making them resistance to such things as insects, chemicals, and drought. Yet, all the ramifications of these transgenic crops cannot be anticipated. For example, when *genetically modified corn* (Bt corn) was introduced in order to make it more insect resistant, the genetic engineers overlooked the possibility that this corn might be harmful to good insects as well, such as the monarch butterfly, which migrates from North to South America annually. Moreover, the fact that corn is a wind pollinated crop was overlooked. Cross pollination resulted which has caused changes in the genetic make up of neighboring natural genetic corn. In addition, the diversity of Mexico's maize crops has been imperiled by wind cross pollination of Bt corn. Will genetically modified corn homogenize Mexico's diverse corn population? Will Ireland's historic potato famine be repeated by a corn famine in Mexico? These are a few of many questions yet to be answered.

smaller replicas. Jordan Pollack at Brandeis University in Walthom, MA, is exploring these properties to design robots that build robots. In addition, biological nanodevices are being studied by Samuel Stupp of Northwestern University. His approach is modeled after how proteins form bones by a self-assembling process. It is hoped this process can be used to manufacture tissues for the pancreas for treatment of diabetes.

nanojets — Michael Moseler and Uzi Landman at Georgia Institute of Technology in Atlanta have been modeling computer simulation

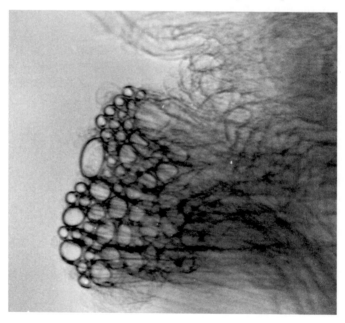

This photograph of nanotubes resembles an abstract drawing. Courtesy of the Smalley Research Group, Rice University, TX.

using the mathematics of hydrodynamic equations. These simulations show the movement of molecules through nanojets nozzles. Researchers plan to do similar simulations using different molecules, such as water, silicon and polymers, to shoot through the nanonozzles. What is their potential use? —ultra small fuel injectors for cleaner engines, nanojets to move genes into cells, and to move and organize nanowires in electronic circuits.

nanolasers — Researchers at the University of California at Berkeley have made a nanolaser using nanowire which can emit a range of ultraviolet light from blue to dark violet with a wavelength of about 17 nanometers. These nanolasers would be useful in identifying chemicals, increasing the memory capacity of disks, and would work with optical computers.

The speed at which the nanoindustries are growing is startling. In 1999 private and venture capital investments were about $100 million and in 2002 it is projected to reach $1 billion, according to a survey by NanoBusiness Alliance, NY. Surf the Internet and discover projects and businesses underway worldwide. There is excitement and promise in the cutting edge world of nanotechnology, but all new scientific ideas must proceed with caution. History has witnessed a number of innovations which have produced global side effects. For example, CPC coolants have been shown to have adverse effects on the ozone layer. When the pesticide DDT was introduced, it was a wonder for controlling insect pests. But, in 1972 the United States banned its use because researchers linked it to the decline of certain bird populations, such as the brown pelican and the bald eagle. Today *genetically modified corn* has been found to adversely impact the diversity of corn crops and beneficial insects. In general, wide spread use of pesticides has created pest resistant strains, and these substances often find their way into the water table. As difficult as it is to create new and useful scientific discoveries, it is more difficult to know all its possible ramifications.

What's to come?

Imagine nanocomputers guiding assembly lines of trillions of nanounits, nanomachines that would be able to self-assemble and self-replicate, fractal shifting robots that could morph their shapes and uses to tackle any task at hand by simply changing their software. *What type of work would nanounits do?* — *clean* toxic wastes (biological units have already been used in this area) and pollution, *deliver* drugs directly to diseased cells, *repair* injured cells, *garden* using nanounits to clear a yard of weeds and prepare enriched soil— *housekeeping* using nano units to dust and clean — *manufacture* foods by assembling ingredients from their atomic state, as in Star Trek's replicator — *recycle* waste and garbage by returning junk to its atoms and molecular states to be used later on to make other products. *Would this mean there would be no famine, no disease, no shortages, no problems?* Maybe. But, can't these nanocomputers have nanoglitches, which may be thought of as nanomutations? *Is nanoworld as small as it gets or can get?* No. There is talk of picotechnology. But perhaps the world of superstrings will scale down nanounits and picounits even further by using strings as building blocks of matter— strings compressed in multiple dimensions whose essence is not defined by atoms but by their vibrations. We can be certain that nanoworld will definitely be one of the steps to the world of the future.

---

[1] *Dialogues Concerning Two New Sciences* (1638). Henry Crew & Alfonso De Slav translators. Macmillan, 1914.

[2] *The nanotube computer* by David Rotman. Technology Review magazine March 2002 issue, Cambridge, MA.

[3] *Travels to the Nanoworld* by Michael Gross, Plenum Trade, NY 1999.

share the work
distributed computing

S uppose you were given one-hundred 4-digit numbers to add up using only pencil and paper. It would be much easier and quicker to divide this task between 20 people who would work simultaneously. That is essentially how *distributed computing* works. A task is divided in to smaller problems and distributed among a system of personal computers, which gathers the results and produces the solution(s). This sort of division of labor works best with problems involving vast quantities of numbers and data crunching.

One of the best distributed computing quests is the search for larger and larger prime numbers. Until the mid 1990s the search for primes was a job tackled primarily by Cray supercomputers — computers which housed thousands of Pentium Pro processors (the Janus, for example, has over 9000 Pentium chips). A personal computer has only one Pentium chip, and consequently is no match for the supercomputer. Yet in 1996, the newest explicit largest prime number was found by using a collection of PCs. At this time a group of individuals linked their computers via the internet. By developing some clever programming, this group unleashed the potential power of a couple of thousand PCs. The success of this computer consortium illustrated the potential use and possibilities of distributed computing. It became clear that the unused power of a PC could be harnessed and connected with other PCs.

Most personal computer power goes unused, even when the computer is being used. A CPU (central processing unit) can perform 100,000,000 instructions per second. If one were able to type 20 keystrokes per second, imagine the unused CPU cycles (CPU cycle is the time needed to get and execute a single machine instruction) spinning around trying to find what work it should be doing while you are frantically typing at a rate that is very sluggish when compared to your computer's potential. Thus a vast number of CPU cycles go unused. It's a waste of computer power and energy. Here's where distributed computing enters the picture. If only 10,000 PCs are linked that would be more Pentium power than a Janus supercomputer.

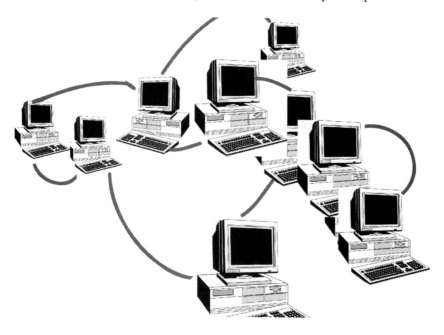

How can you donate your unused CPU cycles and join your computer power with others? The software is all in place, and in fact, today there are tens of thousands of computer users donating their idle computer power to projects and studies throughout the world.

Your computer can participate in problem solving when you're not using it or all of its potential. Some methods work in a similar way as a screen saver by switching on when you're not using your computer and off when you begin using it. The technology is being continually improved. Instead of being confined to a closed commune of PC owners, now the potential computational labor force of computers can be harnessed when connected via the Internet, in essence creating a super *virtual computer.*

What type of problems are being tackled with distributed computing?

- *Protein folding* — Proteins appear as long chains of amino acids responsible for biochemical reactions. They are in essence the cell's work force. As enzymes they take care of the biochemical reactions. As antibodies they help the immune system. Proteins do their work through a *self-assembling* process called *folding.* Being able to simulate protein folding may help scientists discover how to synthetically make polymers, and discover what diseases result when the folding process goes haywire. *Folding@home* is a public service project started by Vijay Pande and his colleagues at Stanford University. Its sister project is *Genome@home.* Their software carries out its calculations whether you're using your computer or or not by recycling the

For protein to carry out the various tasks, they must assume a specific shape called a *fold.* They produce a particular fold by bringing themselves together in order to self-assemble, known as *folding.* This process is carried out in a fraction of a microsecond.

When protein folding goes amiss, their misfolding creates clusters. These clusters account for various diseases such as inherited emphysema, cystic fibrosis, and various cancers. It is felt that when these clusters settle in the brain they account for such diseases as Alzheimer's and Mad Cow.

It is hoped that studying the folding and misfolding process of proteins will help to develop a small molecule that will lead to synthesizing drugs that can either stabilize the folding or disrupt the pathway of the mis-folding protein.

computer's unused CPU cycles. While you're working at your computer, an adjustable small window displays the protein sequence on which your computer is working. The *Folding@home* project's goal is to get 100,000 PC donors involved to work on protein folding. They hope to uncover a protein structure and function, and thereby help design treatments for such diseases as Alzheimer's and Mad Cow diseases. The power of 100,000 PC would be equivalent to a computer with the power of 100,000 Pentium chips, which can overpower today's supercomputers.

• *AIDS research* — The problem results of the distributed computing project *Fight AIDS@home* may help design new drugs for the treatment if AIDS.

• *In search of alien life* — Since May 1999, *SETI@home* has enlisted 2.4 million donors to help search the universe for that special radio signal that will mean extraterrestrial life is trying to make contact.

• *Number theory problems* — These include such problems as helping to factor huge numbers, finding primes nearly a million digits long, joining the GIMPS (Great Internet Mersenne Prime Search).

• *Searching for optimal Golomb rulers* — In the 1960s mathematician Solomon Golomb described *a ruler on which no two pairs of marks measure the same distance.* Such a ruler is called a Golomb ruler. On traditional rulers each pair of marks is the same distance apart. But look at the ruler with the marks at 0, 1, 4, and 6. Checking the distances between all pairs of marks, you see that the distances 1,2,3,4,5,6 are all represented. For a ruler with marks at 0, 1, 3, and 7, its pairs give distances 1,2,3,4,6,7. Distributed computing is used to find and prove *optimal Golomb rulers* — the shortest Golomb ruler for a given number of marks. Is the 0, 1, 4, 6 an optimal Golomb ruler? *Why would one want to find Golomb rulers or care if it is optimal?* Since Golomb rulers deal with various ways to space marks, finding Golomb rulers and deciding if they are optimal is of value in — the science of radio astronomy in which antennas are placed on Golomb marks for the receiving and sending of signals — sensor placements in x-ray crystallography — data encryption — coding theory — combinatorics —communications, and other fields.

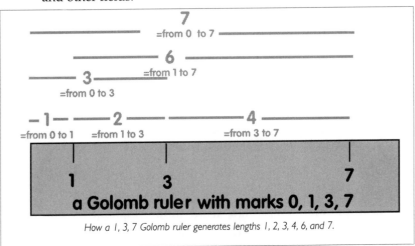

How a 1, 3, 7 Golomb ruler generates lengths 1, 2, 3, 4, 6, and 7.

- analyzing vast amounts of information such as that required in seismic analysis.

- creating models for environmental studies.

- devising simulations for scientific and entertainment uses. For example, *Distributed Science* has a project underway called *GammaFlux,* coordinated by a Swiss nuclear science student, to find the most effective and efficient way to store nuclear waste by modeling gamma radiation escaping from various types of storage containers of nuclear waste.

- predicting financial and economic outcomes.

- analyzing turbulent fluid flow.

- creating lattice models in quantum physics.

- working on computational chemistry and health problems. For example, one project coordinated by *Popular Power* deals with influenza research. Simulations are created to test the effectiveness of vaccines on various influenza virus and mutant strains.

- organizing and distributing music files. An example is P2P technology which links multiple computers to solve various computer problems in computation, collaboration and communication. The music file sharing service, Napster, was one of the best known examples of P2P computing.

- screening compounds as possible cancer drugs.

- mustering 1.4 million PCs in over 200 countries to help screen 3.5 billion molecules to pinpoint anthrax scares

Ever since distributed computing became feasible, the list of i projects it is tackling constantly growing. The business sectors are exploring tapping into distributing computing by offering monetary incentives to those willing to share their CPU cycles. Some people are even predicting that CPU futures may eventually be traded in the marketplace. *Distributed Science* and *Popular Power* are hoping to entice donors by coordinating socially relevant projects and by working out a means to compensate the participating users according to the amount of CPU time contributed.

The Internet has a far reaching and diverse list of distributed computing projects underway in which you can participate and donate your idle CPU cycles. If you are concerned about safety of your files and computer, contact the project that interests you and learn how they have set up their project and the safe guards that are in place.

BIBLIOGRAPHY

Aldridge, S. *The Thread of Life,* Cambridge University Press, Cambridge, 1996.

Auburn, D. *Proof,* Faber & Faber, New York, 2001.

Baker, David R. *First Weave,* San Francisco Chronicle, Dec. 3, 2001.

Beckmann, Petr. *A History of Pi,* St. Martin's Press, NY, 1971.

Blatner, David. *The Joy of Pi,* Walker Publishing Co., 1997.

Briggs, John & Peat, F. David. *Turbulent Mirror,* Harper & Row, New York, 1990.

Brizio, A.M., Brugnoli, M.V., Chastel, A. *Leonardo the Artist,* McGraw-Hill Book Co., New York, 1980.

Cipra, Barry. *What's Happening in Mathematical Sciences 1993,* 1994, 1998-99, American Mathematics Society, Providence, RI, 1993, 1994, 1999.

Corsi, J.R. *Leonardo da Vinci,* Pomegrante Artbooks, Rohnert Park, CA 1995.

Cowen, Ron. *Mining the Skies,* Science News Magazine, vol.159, February 24, 2001.

Davis, J. *Mapping the Code,* John Wiley & Son, Inc., New York, 1990.

Dewdney, A.K. *The Armchair Universe,* W.H. Freeman & Co., New York, 1988.

Dudley, W. editor. *Genetic Engineering,* Greenhaven Press, inc., San Diego, CA, 1990

Dunham, W. *Journey Through Genius,* John Wiley 7 Sons, Inc., New York, 1990.

Engel, Peter. *Folding the Universe,* Vintage Books, NY, 1989.

Garcia, Linda. *The Fractal Explorer,* Dynamic Press, Santa Cruz, CA, 1991.

Garland, T., Kahn, C. *Math and Music,* Dale Seymour Publications, Palo Alto, CA, 1995.

Gelb, M. *How You Can Think Like Leonardo da Vinci,* Dell Trade Paperback, New York, 1998.

Gleick, J. *Chaos,* Penguin Books, New York, 1987.

Goldsworth, A. *Stone,* Harry N. Abrams, Inc., 1994.

Goldsworthy, A. & Craig, D. *Arch,* Harry N. Abrams, Inc., 1999.

Golob, R. & Brus, E. *The Almanac of Science & Technology,* Harcourt Brace Jovanovich, Publishers, New York, 1990.

Gonick, L. & Wheels, M. *The Cartoon Guide to Genetics*, HaperCollins Publishers, New York, 1991.

Goodwin, B. *How the Leopard Got Its Spots*, Simon & Schuster, New York, 1996.

Gross, Micahel. *Travels to the Nanoworld*, Plenum Trade, NY, 1999.

Hellemans, A. & Bryan, B. *The Timetables of Science*, Simon & Schuster, NY, 1988.

Henig, Robin Marantz. *The Monk in the Garden*, Houghton Mifflin Co., 2000.

Hesman, Tina. *The Meaning of Life*, Science News Magazine, vol.157, April 29, 2000.

Hesman, Tina. *Code Breakers*, Science News Magazine, vol.157, June 3, 2000.

Heydenreich, L., Dibner, B., Reti, L. *Leonardo The Inventor*, McGraw-Hill Book Co., New York, 1980.

Jackson, J.F. *Genetics and You*, Humana Press, Totowa, NJ, 1996.

Kappraff, Jay. *Connections, The Geometric Bridge Bewteen Art & Science*, McGraw-Hill, Inc., NY, 1991.

Leiner, Barry M., Cerf, Vinton G., et al. *A Brief History of the Internet*, version3.31, http://www..isoc.org/internet/history.

Mandelbrot, Benoit B. *The Fractal Geometry of Nature*, W.H. Freeman & Co., NY, 1983.

Martin, S. *Picasso at the Lapin Agile*, Gross Press, New York, 1996.

Ochert, Ayala. *Transposons*, Discover Magazine, December 1999.

Palfreman, Jon & Swade, Doron. *The Dream Machines*, BBC Books, London, 1991.

Pappas, Theoni. *The Joy of Mathematics*, Wide World Publishing, San Carlos, CA, 1989.

Pappas, Theoni. *More Joy of Mathematics*, Wide World Publishing/Tetra, San Carlos, CA 1997.

Pappas, Theoni. *The Magic of Mathematics*, Wide World Publishing/Tetra, San Carlos,CA, 1994.

Pappas, Theoni. *Mathematical Footprints*, Wide World Publishing/Tetra, San Carlos,CA, 1999.

Pappas, Theoni. *Math-A-Day*, Wide World Publishing/Tetra, San Carlos,CA, 1999.

Pauling, L. & Hayward, R. *The Archtecture of Molecules*, W.H. Freeman & Co., San Francisco, 1964.

Peitgen, Heinz-Otto, Jürgens, Hartmut, and Saupe, Dietmar. *Chaos and Fractals*, Springer Verlag, New York, 1992.

Pickover, C.A. *Computers, Pattern, Chaos, and Beauty*, St. Martin's Press, New York,1991.

Pickover, Clifford. *Fractal Connections*, St Marin Press, NY, 1996.

Pierce, John. *The Science of Musical Sound*, W.H. Freeman & Co., NY, 1983.

Rheingold, Howard. *Virtual Reality*, Summit Books, New York, 1991.

Richter, J.P. *The Notebooks of Leonrado da Vinci, vols. 1 &2*, Dover Publications, New York, 1970.

Robbin, Tony. *Fourfield: Computers, Art & the 4th dimension*, Little Brown & Company, Boston, 1992.

Singh, S. *Fermat's Enigma*, Walker & Co., New York, 1997.

Stevens, P.S. *Patterns in Nature*, Little, Brown & Co., Boston, 1974.

Stoppard, T. *Arcadia*, Faber & Faber, London, 1993.

Strohman, Richard. *Beyond Genetic Determinism*, Calfornia Monthly, April 2001.

Tame, David. *The Secret Power of Music*, Destiny Books, Vermont, 1984.

Trachtman, Paul. *Redefining Robots*, Smithsonian Magazine, February 2000.

Tudge, Colin. *The Impact of the Gene*, Hill & Wang, New York, 2000.

U.S. Department of Energy Human Genome Program. http://www.ornl.gov/hgmis.

Wells, David. *The Penguin Dictionary of Curious & Interesting Numbers*, Penguin Books, England, 1987.

Wheelwright, Jeff. *Designer Genes*, Smithsonian Magazine, January 2001.

Wolkomir, Richard. *Charting the Terrain of Touch*, Smithsonian Magazine, June 2000.

Zammatio, C., Marinoni, A., Brizio, A.M. *Leonardo the Scientist*, McGraw-Hill Book Co., New York, 1980.

Mathematics teacher and consultant Theoni Pappas received her B.A. from the University of California at Berkeley in 1966 and her M.A. from Stanford University in 1967. Pappas is committed to demystifying mathematics and to helping eliminate the elitism and fear often associated with it.

Her innovative creations include *The Mathematics Calendar, The Children's Mathematics Calendar, The Mathematics Engagement Calendar, The Math-T-Shirt* ,and *What Do You See?*—an optical illusion slide show with text.

Pappas is also the author of the following books: *Mathematics Appreciation, The Joy of Mathematics, Greek Cooking for Everyone, Math Talk, More Joy of Mathematics, Fractals, Googols and Other Mathematical Tales, The Magic of Mathematics, The Music of Reason, Mathematical Scandals, The Adventures of Penrose —The Mathematical Cat, Math for Kids & Other People Too!, and Math-A-Day, Mathematical Footprints. Math Stuff* is her most recent book.

Mathematics Titles by Theoni Pappas

MATH STUFF
$12.95 • 156 pages • illustrated•ISBN:1-884550-26-6

MATHEMATICAL FOOTPRINTS
$10.95 • 156 pages • illustrated•ISBN:1-884550-21-5

MATH-A-DAY
$12.95 • 256 pages • illustrated•ISBN:1-884550-20-7

MATHEMATICAL SCANDALS
$10.95 • 160 pages •illustrated•ISBN:1-884550-10-X

THE MAGIC OF MATHEMATICS
$12.95 • 336 pages •illustrated•ISBN:0-933174-99-3

FRACTALS, GOOGOL, and Other Mathematical Tales?
$9.95 • 64 pages • for all ages • illustrated•ISBN:0-933174-89-6

MATH FOR KIDS & Other People Too!
$10.95 • 140 pages • illustrated•ISBN:1-884550-13-4

ADVENTURES OF PENROSE—The Mathematical Cat
$10.95 • 128 pages • illustrated • ISBN:1-884550-14-2

THE JOY OF MATHEMATICS
$10.95 • 256 pages • illustrated•ISBN:0-933174-65-9

MORE JOY OF MATHEMATICS
$10.95 • 306 pages • illustrated•ISBN:0-933174-73-X
cross indexed with The Joy of Mathematics

MUSIC OF REASON
Experience The Beauty Of Mathematics Through Quotations
$9.95 • 144 pages • illustrated•ISBN:1-884550-04-5

MATHEMATICS APPRECIATION
$10.95 • 156 pages • illustrated•ISBN:0-933174-28-4

MATH TALK
mathematical ideas in poems for two voices
$8.95 • 72 pages • illustrated•ISBN:0-933174-74-8

THE MATHEMATICS CALENDAR
$10.95 • 32 pages • written annually • illustrated • ISBN:1-884550-

THE CHILDREN'S MATHEMATICS CALENDAR
$10.95 • 32pages • written annually • illustrated • ISBN:1-884550-

WHAT DO YOU SEE?
An Optical Illusion Slide Show with Text
$29.95 • 40 slides • 32 pages illustrated•ISBN:0-933174-78-0